MATHEMATICS

GCSE Grade Booster

B. K. Holdsworth

Schofield & Sims Ltd.

0 7217 4611 x

First printed 1989

Schofield & Sims Ltd.
Dogley Mill
Fenay Bridge
Huddersfield
HD8 0NQ
England

Designed by Graphic Art Concepts, Leeds

Printed in Great Britain at the Alden Press, Oxford

Contents

Introduction

This Mathematics Grade Booster has been designed to help you revise the core elements of the GCSE course. The book contains the essential information for each Level (Foundation, Intermediate and Higher), all clearly labelled. It does not reproduce the Educational Aims for following a course in Mathematics, provided by each Examination Board; nor does it give the schemes for assessment and the assessment objectives. You should ask your teacher for these.

The book has been divided into 21 Modules, each sub-divided into up to three parts. If you are concerned only with the Foundation Level, then you need read and know only the content included in the relevant part of each Module. If you are concerned with the Intermediate Level, you must read and learn everything in the appropriate two parts of each Module. There is no time-limit for each Module – you should proceed to the next one only when you are convinced that you are ready. For the Higher Level, of course, you will need to know and understand everything in all Modules.

Calculators: Foundation Level

You will need a calculator that has at least the five keys for addition, subtraction, multiplication, division and square root ($+, -, \times, \div$ and $\sqrt{}$). Since the layout style varies from calculator to calculator, you should investigate what your calculator can do and how it does the operations. Use the instruction guide to help you.

Calculators: Intermediate Level

Calculators use *standard form* for writing very large or very small numbers. Any number can be written in the form $A \times 10^n$ where n is an integer and A is a number between 1 and 10 or equal to 1, *i.e.* $1 \leqslant A < 10$.

Example:

$53 = 5.3 \times 10^1$	$0.1 = 1 \times 10^{-1}$
$765371 = 7.65371 \times 10^5$	$0.0000086 = 8.6 \times 10^{-6}$
$0.006 = 6 \times 10^{-3}$	$102 = 1.02 \times 10^2$

If you see 3.49325 09 on the calculator, this means 3.49325×10^9
$= 3\,493\,250\,000$

If you see 1.4976 –08 on the calculator, this means 1.4976×10^{-8}
$= 0.000000014976$

1 Types of Number

Foundation Level
Whole numbers, odd and even numbers, prime numbers, square numbers, cubic numbers, factors, multiples, square root, number sequence, number patterns.

Intermediate Level
Integers, rational numbers, irrational numbers, square roots, highest common factor (HCF), lowest common multiple (LCM), indices, order of operations, ideas of ordering.

Higher Level
Fractional indices.

Foundation Level

Whole numbers

These are often called *counting numbers* or *natural numbers:*

$$\{1, 2, 3, 4, 5, \ldots\}.$$

Odd numbers

These are whole numbers which will not divide by 2 to give another whole number exactly:

$$\{1, 3, 5, \ldots\}.$$

Even numbers

These are whole numbers which will divide by 2 to give another whole number exactly:

$$\{2, 4, 6, \ldots\}.$$

Prime numbers

A prime number is a whole number which can only be divided by itself and 1 to give another whole number. Since 1 divides into any number to give the number itself, 1 is unique and is therefore *not* included in the set of prime numbers:

$$\{2, 3, 5, 7, 11, 13, \ldots\}.$$

Square numbers

These are formed when any whole number is multiplied by itself:

$$3 \times 3 = 9 \quad \text{or} \quad 7 \times 7 = 49 \quad \text{or} \quad 9 \times 9 = 81$$

i.e. 9, 49 and 81 are examples of square numbers.

Mathematicians use a 'shorthand' for writing 3×3. They write 3^2. This is read as '3 to the power of 2' or '3 squared'.

The set of square numbers is therefore

$$\{1^2, 2^2, 3^2, 4^2, \ldots\} \quad \text{or} \quad \{1, 4, 9, 16, \ldots\}.$$

Cubic numbers

These are formed when any whole number is multiplied by itself and again by itself.

$3 \times 3 \times 3 = 27$, or $4 \times 4 \times 4 = 64$, or $9 \times 9 \times 9 = 729$

i.e. 27, 64, 729 are examples of cubic numbers.

Again, a shorthand is used for writing $9 \times 9 \times 9$. It is 9^3. This is read as '9 to the power of 3' or '9 cubed'.

The set of cubic numbers is therefore

$\{1^3, 2^3, 3^3, 4^3, \ldots\}$ or $\{1, 8, 27, 64, \ldots\}$.

Factors of whole numbers

Whenever two whole numbers are multiplied together to give some other whole number, either of the two numbers multiplied is said to be a factor of the answer,

i.e. $12 = 3 \times 4$, then 3 is a factor of 12
4 is also a factor of 12.

$12 = 2 \times 6$, then 2 and 6 are both factors of 12.

$12 = 1 \times 12$, so of course 1 and 12 are both factors of 12.

Multiples of whole numbers

Whenever a whole number is divided exactly by another whole number, so that there is no remainder, the first whole number is said to be a multiple of the second whole number,

i.e. $\frac{20}{5} = 4$, then 20 is a multiple of 5

or $\frac{20}{4} = 5$, then 20 is a multiple of 4.

Square root

When two identical numbers are multiplied together, one of the numbers is said to be the square root of the answer to the multiplication,

i.e. $3 \times 3 = 9$, then 3 is the square root of 9

$5 \times 5 = 25$, then 5 is the square root of 25.

Again, mathematicians use a shorthand for 'the square root of'.

The symbol is $\sqrt{\ }$ or $\sqrt[2]{\ }$, *i.e.* $3 = \sqrt{9}$ or $3 = \sqrt[2]{9}$.

Usually the first notation is used

so $5 = \sqrt{25}$.

Number sequence

This is a set of numbers in which the next term is obtained by inspecting the given set of numbers and noting the rule that connects them.

Example 1: 2, 5, 8, 11, . . .

The next term must be 14, since each term differs by 3.

Example 2: 3, 6, 12, 24, . . .

The next term must be 48, since each term can be obtained by multiplying the previous term by 2.

Number patterns

Some numbers can be represented by dots to form particular patterns.

Example 1: Square numbers {1, 4, 9, 16, 25, . . .}.

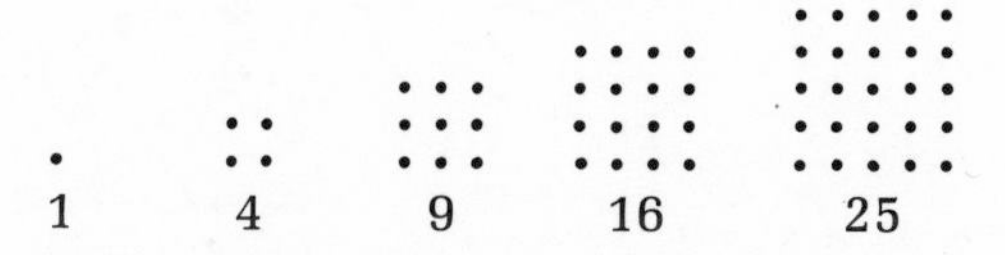

1 4 9 16 25

Example 2: Triangular numbers {1, 3, 6, 10, . . .}.

1 3 6 10

What is the next triangular number?

It can be shown to be 15 by the dot pattern, or by inspecting the sequence:

1 difference of 2 3 difference of 3 6 difference of 4 10 difference of 5

NOTE: The sum of two adjacent triangular numbers always gives a square number.

$1 + 3 = 4$

or

$6 + 10 = 16$

Intermediate Level

Integers

An integer is a member of the set of positive and negative whole numbers including zero:

$\{\ldots, -3, -2, -1, 0, 1, 2, 3, \ldots\}$

sometimes written as

$\{\ldots, {}^{-}3, {}^{-}2, {}^{-}1, 0, 1, 2, 3, \ldots\}$.

Rational numbers

These are numbers which can be written as a ratio of two integers (*not* including the case where the divisor is zero). Rational numbers are fractions.

$\{\frac{1}{2}, -\frac{3}{5}, 0.7, \frac{5}{3}, \ldots\}$. *N.B.* $0.7 = \frac{7}{10}$

Irrational numbers

These are numbers which cannot be written as the ratio of two integers. Many numbers cannot be written as an exact fraction, common examples being π (used in measurement of circles), sin 21° (see trigonometry later), $\sqrt{2}$ and $\sqrt{7}$. These irrational numbers cannot be written as an exact decimal either, as you will see if you try them on your calculator.

Square roots

Most square roots are irrational.

Example: $\sqrt{7} \simeq 2.6457513$

Note that this is only an approximate number. The degree of accuracy will vary (see Module 5).

Highest common factor (HCF)

The HCF of two or more numbers is the largest whole number that can be divided into each of the two or more numbers without leaving a remainder.

Example: Find the HCF of 36 and 90.

$36 = 2 \times 18 = 2 \times 2 \times 9 = 2 \times 2 \times 3 \times 3$

$90 = 2 \times 45 = 2 \times 3 \times 15 = 2 \times 3 \times 3 \times 5$

$\text{HCF} = 2 \times 3 \times 3 = 18$

Express the given numbers in terms of prime factors, take the factors which are common and multiply these to get the HCF.

Lowest common multiple (LCM)

The LCM of two or more whole numbers is the smallest number into which each of the two or more whole numbers is exactly divisible.

Example: Find the LCM of 36 and 90.

$36 = 2 \times 18 = 2 \times 2 \times 9 = 2 \times 2 \times 3 \times 3$

$90 = 2 \times 45 = 2 \times 3 \times 15 = 2 \times 3 \times 3 \times 5$

$\text{LCM} = 2 \times 2 \times 3 \times 3 \times 5 = 180$

Express the given numbers in terms of prime factors, take all the factors of the first number and the factors of the second number that are not factors of the first number and multiply these to get the LCM.

Indices

We noted earlier that the shorthand for 3×3 is 3^2, and that $9 \times 9 \times 9$ may be written 9^3. The ^{2} in 3^2 or ^{3} in 9^3 are called *indices* (or exponents). The singular term is an *index* (or exponent).

Rules of indices

$$5 \times 5 = 5^2 \qquad 5 \times 5 \times 5 \times 5 = 5^4$$
$$5 \times 5 \quad \times \quad 5 \times 5 \times 5 \times 5 = 5^6$$
$$\text{or} \quad 5^2 \quad \times \quad 5^4 \quad = 5^6$$

When *multiplying* like bases (base 5 above), *add* the indices. In general, if the base is a then $a^p \times a^q = a^{p+q}$ where p and q are indices.

$$5^6 \div 5^4 = \frac{\cancel{5}^1 \times \cancel{5}^1 \times \cancel{5}^1 \times \cancel{5}^1 \times 5 \times 5}{\cancel{5}_1 \times \cancel{5}_1 \times \cancel{5}_1 \times \cancel{5}_1} = 5^2$$

$$\text{or} \quad 5^6 \div 5^4 = 5^{6-4} = 5^2$$

When *dividing* like bases, *subtract* the indices,

$$\text{i.e.} \quad \frac{a^p}{a^q} = a^{p-q}$$

N.B. $\dfrac{5^4}{5^6} = 5^{-2}$ or $\dfrac{\cancel{5}^1 \times \cancel{5}^1 \times \cancel{5}^1 \times \cancel{5}^1}{\cancel{5}_1 \times \cancel{5}_1 \times \cancel{5}_1 \times \cancel{5}_1 \times 5 \times 5} = \dfrac{1}{5^2}$

$$\therefore 5^{-2} = \frac{1}{5^2}$$

In general $a^{-p} = \dfrac{1}{a^p}$.

$$(5^2)^4 = 5^2 \times 5^2 \times 5^2 \times 5^2 = 5^8.$$

In general $(a^p)^q = a^{p \times q}$.

Order of operations

This is sometimes called the Rule of Precedence. We must adopt some rules when dealing with operations of addition, subtraction, multiplication and division. These rules are *universal* (always to be followed). The order of operations is

1st evaluate brackets
2nd deal with division and multiplication before
3rd addition and subtraction.

NOTE: The mnemonic **BODMAS** might help (**B**rackets **O**f **D**ivision), **M**ultiplication, **A**ddition, **S**ubtraction) for the order of operations.

Example 1: $6 - 10 \div 5$ (Division must be done before subtraction)
$= 6 - 2$
$= 4$

Example 2: $26 - (15 - 3)$ (Evaluate bracket first)
$= 26 - 12$
$= 14$

Example 3: $12 - 2 \times 3 + 6$ (Evaluate 2×3 first)
$= 12 - 6 + 6$
$= 12$

Example 4: $(8 + 10) \div (6 - 3)$ (Evaluate both sets of brackets)
$= 18 \div 3$
$= 6$

Ideas of ordering

We are all familiar with the equality sign (=).

In mathematics, just as in everyday life, many things are not equal and we have symbols (again, the use of a shorthand) to denote them:

$\neq$ is not equal to (a slant line through any symbol negates that symbol)

$>$ is greater than $\geqslant$ is greater than or equal to

$<$ is less than $\leqslant$ is less than or equal to

Examples: $1 + 2 = 3$ $1 + 2 \neq 4$ $6 > 3$ $5 < 8$

Higher Level

Fractional Indices

$x^{\frac{1}{2}} \times x^{\frac{1}{2}} = x^1$ (Index rule)

$7 \times 7 = 49$ 7 is the square root of 49.

$10 \times 10 = 100$ 10 is the square root of 100.

$\therefore x^{\frac{1}{2}}$ is the square root of x, also written $\sqrt[2]{x}$ or simply $\sqrt{x}$.

Similarly $x^{\frac{1}{3}} \times x^{\frac{1}{3}} \times x^{\frac{1}{3}} = x^1$

$\therefore x^{\frac{1}{3}}$ is the cube root of x, also written $\sqrt[3]{x}$

Likewise $x^{\frac{3}{4}} \times x^{\frac{3}{4}} \times x^{\frac{3}{4}} \times x^{\frac{3}{4}} = x^3$

$\therefore x^{\frac{3}{4}}$ is the fourth root of x^3, also written $\sqrt[4]{x^3}$

In general: $x^{\frac{p}{q}}$ is the qth root of x^p, also written $\sqrt[q]{x^p}$

$x^{-\frac{1}{2}} = \frac{1}{x^{\frac{1}{2}}}$ or $\frac{1}{\sqrt{x}}$ Remember $a^{-p} = \frac{1}{a^p}$

$x^{-\frac{1}{3}} = \frac{1}{x^{\frac{1}{3}}}$ or $\frac{1}{\sqrt[3]{x}}$

Example 1: $5^{-\frac{1}{2}} \times 5^{\frac{3}{2}} = 5^{-\frac{1}{2}+\frac{3}{2}} = 5^1 = 5$

Example 2: $\frac{6^{\frac{4}{5}}}{6^{-\frac{2}{5}}} = 6^{\frac{4}{5}} \times 6^{\frac{2}{5}} = 6^{\frac{6}{5}}$ or $\sqrt[5]{6^6}$

2 Fractions and Percentages

Foundation Level

Fractions, vulgar fractions, improper fractions, mixed numbers, equivalent fractions, ordering of fractions, addition and subtraction of fractions, multiplying fractions, division of fractions, operations with fractions, decimals, adding and subtracting decimals, multiplying decimals, division, changing vulgar and improper fractions to decimals, percentages.

Intermediate Level

Changing decimals to fractions, conversion to percentages, percentage change.

Higher Level

Reverse percentages.

Foundation Level

Fractions

A fraction is a part of a whole.

The strip represents a whole.
The strip is divided into 4 equal parts.

The shaded portion is $\frac{1}{4}$ of the strip: this is said to be one-fourth or a quarter of the strip.

The number 4 below the line in $\frac{1}{4}$ is called the *denominator* of the fraction and shows how many equal parts there are in the whole. The number above the line is called the *numerator* of the fraction and indicates how many parts are taken to represent the complete fraction.

Example: $\frac{3}{8}$ means the whole has been divided into 8 equal parts and that 3 of these equal parts are being considered.

Vulgar fractions

Whenever the numerator of a fraction is a number less than that of the denominator, the fraction is said to be a vulgar fraction or a proper fraction.

Examples: $\frac{1}{4}$, $\frac{3}{8}$, $\frac{6}{7}$, $\frac{9}{10}$.

Some texts show them like this: 1/4, 3/8, 6/7, 9/10.

Improper fractions

This is the converse of a proper fraction. Therefore whenever the numerator of a fraction is greater than that of the denominator, the fraction is said to be an improper fraction.

Examples: $\frac{5}{4}$, $\frac{7}{5}$, $\frac{12}{7}$, $\frac{71}{10}$.

Mixed numbers

Every improper fraction can be expressed as a whole number and a proper fraction. This is called a mixed number.

Examples: $\frac{5}{4} = 1 + \frac{1}{4} = 1\frac{1}{4}$; $\quad \frac{7}{5} = 1 + \frac{2}{5} = 1\frac{2}{5}$;

$\frac{12}{7} = 1 + \frac{5}{7} = 1\frac{5}{7}$; $\quad \frac{71}{10} = 7 + \frac{1}{10} = 7\frac{1}{10}$

In each case, see how many wholes can be obtained by division. The remainder is then expressed as a proper fraction.

Equivalent fractions

If we consider a whole and divide it into 4 equal parts and shade 3 parts of it, the shaded portion is $\frac{3}{4}$.

Example:

If we take the same whole and divide it into 8 equal parts and shade 6, the shaded portion is $\frac{6}{8}$.

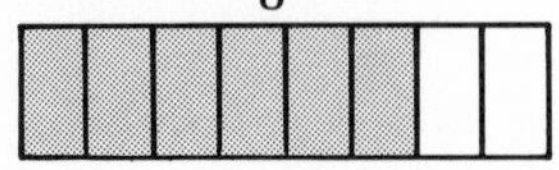

It can be clearly seen that $\frac{3}{4} = \frac{6}{8}$. These are said to be equivalent fractions.

The value of any fraction is unchanged if both the numerator and denominator are multiplied or divided by the same number.

Examples: $\frac{3}{4} = \frac{3 \times 2}{4 \times 2} = \frac{6}{8}$; $\quad \frac{3}{4} = \frac{3 \times 5}{4 \times 5} = \frac{15}{20}$;

$\frac{36}{48} = \frac{36 \div 12}{48 \div 12} = \frac{3}{4}$; $\quad \frac{120}{160} = \frac{120 \div 40}{160 \div 40} = \frac{3}{4}$

There are many fractions equivalent to $\frac{3}{4}$. Four of them are $\frac{6}{8}$, $\frac{15}{20}$, $\frac{36}{48}$, $\frac{120}{160}$ but each is $\frac{3}{4}$ in the *lowest form*.

Fractions such as $\frac{3}{4}$, $\frac{5}{8}$, $\frac{7}{10}$, $\frac{9}{11}$ are said to be in their lowest form. It is normal practice to express any fractional answer in its lowest form.

Examples: $\frac{9}{24} = \frac{3}{8}$ (dividing numerator and denominator by the same number, 3)

$\frac{14}{16} = \frac{7}{8}$ (dividing numerator and denominator by 2)

$\frac{33}{44} = \frac{3}{4}$ (dividing numerator and denominator by 11)

Examples such as the above are said to be 'cancelled down'.

Ordering of fractions

This is done by making equivalent fractions of a set of fractions, with the same denominator.

Example: Arrange $\frac{3}{5}, \frac{3}{4}, \frac{2}{3}, \frac{7}{12}$ in ascending order starting with the smallest.

We need to find a denominator which is common to 5, 4, 3 and 12. In fact, this is the LCM (Lowest Common Multiple) defined in Module 1.

$$\left.\begin{array}{r} 5 = 5 \times 1 \\ 4 = 2 \times 2 \\ 3 = 3 \times 1 \\ 12 = 2 \times 2 \times 3 \end{array}\right\} \quad \text{LCM} = 5 \times 2 \times 2 \times 3 = 60$$

Then convert to equivalent fractions:

$$\frac{3}{5} = \frac{3 \times 12}{5 \times 12} = \frac{36}{60}; \qquad \frac{3}{4} = \frac{3 \times 15}{4 \times 15} = \frac{45}{60};$$

$$\frac{2}{3} = \frac{2 \times 20}{3 \times 20} = \frac{40}{60}; \qquad \frac{7}{12} = \frac{7 \times 5}{12 \times 5} = \frac{35}{60}$$

Hence the ascending order is $\frac{35}{60}, \frac{36}{60}, \frac{40}{60}, \frac{45}{60}$

$\therefore$ order is $\frac{7}{12}, \frac{3}{5}, \frac{2}{3}, \frac{3}{4}$

If the question had required the answer in descending order, then it would obviously be $\frac{3}{4}, \frac{2}{3}, \frac{3}{5}, \frac{7}{12}$.

Addition and subtraction of fractions

Only fractions having the same denominators can be added or subtracted. Therefore we need to find the lowest common denominator again in order to form equivalent fractions.

The method for *addition* is as follows.

Example 1: $\frac{1}{4} + \frac{2}{3}$

Find the lowest common denominator (i.e. the LCM of 4 and 3) which is $2 \times 2 \times 3 = 12$.

$$\therefore \frac{1}{4} = \frac{1 \times 3}{4 \times 3} = \frac{3}{12}; \qquad \frac{2}{3} = \frac{2 \times 4}{3 \times 4} = \frac{8}{12};$$

$\therefore \frac{1}{4} + \frac{2}{3} = \frac{3}{12} + \frac{8}{12} = \frac{11}{12}$ (Just add the numerators to give the numerator of the new fraction.)

Example 2: $\frac{3}{4} + \frac{5}{6} = \frac{9}{12} + \frac{10}{12} = \frac{19}{12}$

The improper fraction is now written in mixed number form:

$$\frac{19}{12} = 1 + \frac{7}{12} = 1\frac{7}{12}$$

Example 3: $3\frac{1}{2} + 1\frac{1}{6} + 2\frac{3}{4}$

It is usual to add the whole numbers together first and then add the fractional parts.

i.e. $3 + 1 + 2 + \frac{1}{2} + \frac{1}{6} + \frac{3}{4} = 6 + \frac{6}{12} + \frac{2}{12} + \frac{9}{12}$

$= 6\,\frac{17}{12} = 6 + 1 + \frac{5}{12} = 7\frac{5}{12}$

or Some people prefer to convert mixed numbers first to improper fractions.

$3\frac{1}{2} + 1\frac{1}{6} + 2\frac{3}{4} = \frac{7}{2} + \frac{7}{6} + \frac{11}{4}$

$= \frac{42}{12} + \frac{14}{12} + \frac{33}{12} = \frac{89}{12} = 7 + \frac{5}{12} = 7\frac{5}{12}$

You should choose which method you find easier, though the first method is thought to be easier for the addition of fractions.

The method for *subtraction* of fractions is very similar to that used for addition. You must first find the lowest common denominator.

Example 1: $\frac{2}{3} - \frac{3}{8} = \frac{2 \times 8}{3 \times 8} - \frac{3 \times 3}{8 \times 3} = \frac{16}{24} - \frac{9}{24} = \frac{7}{24}$

Example 2: $\frac{5}{9} - \frac{1}{3} = \frac{5}{9} - \frac{3}{9} = \frac{2}{9}$

Example 3: $2\frac{2}{3} - 1\frac{1}{5}$

It is usual to subtract the whole numbers and then subtract the fractions.

$2\frac{2}{3} - 1\frac{1}{5} = 2 - 1 + \frac{2}{3} - \frac{1}{5} = 1 + \frac{10}{15} - \frac{3}{15} = 1 + \frac{7}{15} = 1\frac{7}{15}$

or Convert mixed numbers to improper fractions.

$2\frac{2}{3} - 1\frac{1}{5} = \frac{8}{3} - \frac{6}{5} = \frac{40}{15} - \frac{18}{15} = \frac{22}{15} = 1\frac{7}{15}$

In subtraction, the second method is often more convenient. Consider the following.

Example 4: $5\frac{2}{5} - 3\frac{3}{4}$

Either: $5 - 3 + \frac{2}{5} - \frac{3}{4} = 2 + \frac{8}{20} - \frac{15}{20} = 2 - \frac{7}{20}$

i.e. $1 + 1 - \frac{7}{20} = 1 + \frac{20}{20} - \frac{7}{20} = 1\frac{13}{20}$

or $\frac{27}{5} - \frac{15}{4} = \frac{108}{20} - \frac{75}{20} = \frac{33}{20} = 1\frac{13}{20}$

You need to be careful with the first method, and you may find the second one safer.

Combined addition and subtraction

Example: $2\frac{1}{4} + 3\frac{1}{2} - 1\frac{3}{8} - 1\frac{1}{12} = 2 + 3 - 1 - 1 + \frac{1}{4} + \frac{1}{2} - \frac{3}{8} - \frac{1}{12}$

$= 3 + \frac{6}{24} + \frac{12}{24} - \frac{9}{24} - \frac{2}{24}$

$= 3 + \frac{7}{24} = 3\frac{7}{24}$

Multiplying fractions

Numerators are multiplied together to give the numerator of the new fraction and the denominators are multiplied together to give the denominator of the new fraction. Mixed numbers must always be converted to improper fractions.

Example 1: $\frac{3}{4} \times \frac{5}{7} = \frac{3 \times 5}{4 \times 7} = \frac{15}{28}$

Example 2: Either: $\frac{3}{5} \times 4\frac{1}{6} = \frac{3}{5} \times \frac{25}{6} = \frac{75}{30} = 2 + \frac{15}{30} = 2 + \frac{1}{2} = 2\frac{1}{2}$

(NB $\frac{15}{30} = \frac{1}{2}$ in its lowest terms)

or: $\frac{3}{5} \times 4\frac{1}{6} = \frac{\not{3}^{1}}{{}_{1}\not{5}} \times \frac{\not{25}^{5}}{\not{6}_{2}}$ (cancelling factors) $= \frac{5}{2} = 2\frac{1}{2}$

You must choose which method you prefer. The method of cancelling factors is using the rule for the equivalence of fractions.

Example 3: $\frac{3}{5}$ of 25 means $\frac{3}{5} \times 25$ (replace of by multiplication sign)

or $\frac{3}{5} \times \frac{25}{1}$ (25 wholes is $\frac{25}{1}$) $= \frac{75}{5} = 15$

Division of fractions

Dividing any number by 2 is like multiplying by $\frac{1}{2}$:

$10 \div 2 = 5$ or $\frac{10}{1} \times \frac{1}{2} = \frac{10}{2} = 5.$

Dividing any number by 3 is like multiplying by $\frac{1}{3}$:

$\frac{12}{3} = 4$ or $\frac{12}{1} \times \frac{1}{3} = \frac{12}{3} = 4.$

Dividing any number by $\frac{3}{2}$ is like multiplying by $\frac{2}{3}$:

$6 \div 1\frac{1}{2} = 4$ or $\frac{6}{1} \times \frac{2}{3} = \frac{12}{3} = 4.$

In fact to divide by any fraction, all we need to do is to invert the fraction and multiply by it.

Example 1: $\frac{2}{3} \div \frac{3}{4} = \frac{2}{3} \times \frac{4}{3} = \frac{8}{9}$

Example 2: $\frac{3}{5} \div \frac{3}{8} = \frac{3}{5} \times \frac{8}{3} = \frac{24}{15} = 1\frac{9}{15} = 1\frac{3}{5}$

or cancelling factors $\frac{{}^{1}\not{3}}{5} \times \frac{8}{\not{3}_1} = \frac{8}{5} = 1\frac{3}{5}$

Mixed numbers should always be changed to improper fractions.

Example 3: $6\frac{3}{4} \div 1\frac{1}{2} = \frac{27}{4} \div \frac{3}{2} = \frac{27}{4} \times \frac{2}{3} = \frac{54}{12} = 4\frac{6}{12} = 4\frac{1}{2}$

or $\frac{{}^{9}\not{27}}{{}_{2}\not{4}} \times \frac{\not{2}^{1}}{\not{3}_{1}} = \frac{9}{2} = 4\frac{1}{2}$

Example 4: $2\frac{1}{5} \div 3\frac{3}{10} = \frac{11}{5} \div \frac{33}{10} = \frac{{}^{1}\not{11}}{{}_{1}\not{5}} \times \frac{\not{10}^{2}}{\not{33}_{3}} = \frac{2}{3}$

Operations with fractions

These are done just as for whole numbers.

Example: $\left(2\frac{3}{5} + 1\frac{1}{2}\right) \div \left(\frac{1}{4} \times \frac{82}{3}\right) = \left(3 + \frac{3}{5} + \frac{1}{2}\right) \div \left(\frac{\overset{41}{\not{82}}}{\underset{6}{\not{12}}}\right)$

$$= \left(3 + \frac{6}{10} + \frac{5}{10}\right) \div \left(\frac{41}{6}\right)$$

$$= \left(3 + \frac{11}{10}\right) \div \left(\frac{41}{6}\right)$$

$$= \frac{{}^{1}\not{41}}{{}_{5}\not{10}} \times \frac{\not{6}^{3}}{\not{41}_{1}} = \frac{3}{5}$$

Decimals

These are fractions which have denominators of 10, 100, 1000, 10 000, 100 000, 1 000 000, etc.

Examples: $\frac{3}{10}, \frac{51}{100}, \frac{799}{1000}, \frac{13}{10\,000}, \frac{21}{10}, \frac{7896}{1\,000\,000}$

$2\frac{7}{10}$ is written as 2.7. The dot is called the decimal point and separates whole numbers from the fractional parts.

$$\frac{3}{10} = 0.3; \quad \frac{51}{100} = 0.51; \quad \frac{799}{1000} = 0.799; \quad \frac{13}{10\,000} = 0.0013;$$

$$\frac{21}{10} = 2.1; \quad \frac{7896}{1\,000\,000} = 0.007896$$

or in words $\frac{3}{10} = 0.3$ is three tenths

$\frac{51}{100} = 0.51$ is fifty-one hundredths or 5 tenths and 1 hundredth

$\frac{21}{10} = 2.1$ is 2 wholes and 1 tenth

$\frac{13}{10\,000} = 0.0013$ is 13 ten thousandths

etc.

Adding and subtracting decimals

These operations are done just as for whole numbers. It is helpful to work in columns, keeping the decimal points under each other.

Example 1: Add 10.601, 3.709, 2.55 and 3.108.

```
10.601
 3.709
 2.55
 3.108
------
19.968
```

Example 2: Subtract 25.04 from 119.619

```
119.619
 25.04
-------
 94.579
```

Multiplying decimals

Multiplying any number by 10 moves the digits of the number *one* place to the left:

$25 \times 10 = 250 \qquad 2.5 \times 10 = 25 \qquad 7.62 \times 10 = 76.2$

Multiplication by 100 moves the digits of the number *two* places to the left:

$2.5 \times 100 = 250 \qquad 7.62 \times 100 = 762 \qquad 8.1397 \times 100 = 813.97$

Multiplication by 1000 moves the digits of the number *three* places to the left:

$0.00006179 \times 1000 = 0.06179$ etc.

When multiplying decimals, add the number of decimal places in both decimals being multiplied to find the number of decimal places in the answer.

Example: 35.1×0.072

Ignore initially the decimal points and multiply 351 by 72.

```
  351
×  72
-----
24570
  702
-----
25272
```

Number of decimal places in 35.1×0.072 is 1 in first number, 3 in second number, *i.e.* $1 + 3 = 4$ places required in answer.

So the answer is 2.5272.

Division

Always make the divisor into a whole number and therefore always do the same compensation to the dividend (the number being divided).

Example: $2.04 \div 1.7 = \frac{2.04}{1.7}$

Multiply divisor 1.7 by 10 to make a whole number. Therefore multiply dividend (2.04) also by 10.

Hence $\frac{2.04}{1.7} = \frac{2.04 \times 10}{1.7 \times 10} = \frac{20.4}{17} = 1.2.$

Changing vulgar and improper fractions to decimals
Divide out using short/long division methods, or simply use your calculator.

Example 1: $\frac{1}{4} = 0.25$

Example 2: $\frac{3}{8} = 0.375$

Example 3: $\frac{1}{3} = 0.333333\ldots$
or
$0.\dot{3}$ ← indicates the recurring decimal

Example 4: $2\frac{5}{16} = \frac{37}{16} = 2.3125$

$$\begin{array}{r} 2.3125 \\ 16\overline{)37} \\ \underline{32} \\ 50 \\ \underline{48} \\ 20 \\ \underline{16} \\ 40 \\ \underline{32} \\ 80 \\ \underline{80} \end{array}$$

Percentages
A fraction with a denominator of 100 gives the numerator as a percentage. Percentages means 'parts of a hundred'.
Percentages are useful when comparing situations or when dealing with money problems involving profit and loss, discounts, etc.
25% means twenty-five per cent and represents 25 parts of 100 parts. It is written as $\frac{25}{100} = \frac{1}{4}$.

60% means $\frac{60}{100} = \frac{3}{5}$ 80% means $\frac{80}{100} = \frac{4}{5}$ 12% means $\frac{12}{100} = \frac{3}{25}$

Example 1: Find 15% of £20.

$15\% \text{ of } £20 = \frac{15}{100} \times £20 = £\frac{300}{100} = £3$

or $£\frac{\cancel{15}^{3}}{{}_{1}\cancel{5}\cancel{100}} \times \cancel{20}^{1} = £3$

Example 2: Find 20% of £16.

$20\% \text{ of } £16 = \frac{20}{100} \times £16 = £\frac{320}{100} = £3{\cdot}20$

or $£\frac{\cancel{20}^{1}}{{}_{5}\cancel{100}} \times 16 = £\frac{16}{5} = £3\frac{1}{5} = £3{\cdot}20$

Intermediate Level

Changing decimals to fractions
Remember, decimals are fractions with denominators of 10, 100, 1000, etc.

Example 1: $0.5 = \frac{5}{10} = \frac{1}{2}$

Example 2: $0.15 = \frac{15}{100} = \frac{3}{20}$

Example 3: $0.075 = \frac{75}{1000} = \frac{3}{40}$

Conversion from vulgar fractions and decimals to percentages

Multiply vulgar fraction or decimal by 100 to convert to a percentage.

Example 1: $\frac{5}{10}$ $\frac{5}{10} \times 100\% = \frac{500}{10}\% = 50\%$

Example 2: 0.06 $0.06 \times 100\% = 6\%$

Example 3: 0.075 $0.075 \times 100\% = 7.5\%$

Example 4: $\frac{17}{1000}$ $\frac{17}{1000} \times 100\% = \frac{1700}{1000} = 1.7\%$

Percentage change

This is always found by the formula $\frac{\text{Actual change}}{\text{Original value}} \times 100\%$.

Example 1: Find the % profit when an article costing £20 is sold for £24.

Profit = £(24 − 20) = £4

$\therefore$ % profit $= \frac{4}{20} \times 100\% = \frac{400}{20}\% = 20\%$

Remember, the denominator is the original price of £20.

Example 2: Find the % loss when an article which originally cost £30 is sold for £25.

Loss = £(30 − 25) = £5

% loss $= \frac{5}{30} \times 100\% = \frac{500}{30}\% = 16\frac{2}{3}\%$

Higher Level

Reverse percentages

Example 1: Find cost price of an article that was sold for £50 at a profit of 20% on the cost price.

Let the cost price be 100% (CP)

$\therefore$ selling price is 120% (SP) $\therefore$ 120% of CP is £50

$\therefore$ 1% of CP is $= £\frac{50}{120}$

$\therefore$ 100% of CP (*i.e.* whole CP) $= £\frac{50}{120} \times 100$

$= £\frac{500}{12} = £\frac{125}{3} = £41\frac{2}{3}$

or £41·67 (to nearest penny)

Example 2: A car is sold for £1440 at a loss of 80% on its original value. What was the original price?

Let cost price be 100% (CP)

$\therefore$ selling price is 20% of CP $\therefore$ 20% of CP is £1440

$\therefore$ 1% of CP is $£\frac{1440}{20} = £72$

Original price is 100%, *i.e.* £72 × 100 = £7200.

3 Directed and Negative Numbers

Foundation Level

Directed numbers, adding directed numbers, subtracting directed numbers, multiplication of directed numbers, division of directed numbers.

Foundation Level

Directed numbers

So far we have considered numbers greater than 0 by looking at whole numbers, fractions, etc. All these numbers could be represented by a number line, going either vertically or horizontally.

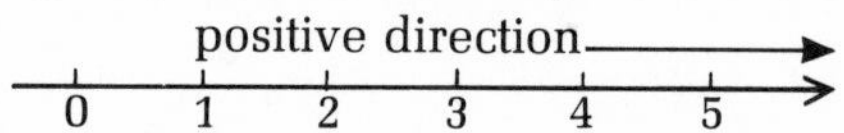

Numbers greater than 0 travel to the right, which we could say is the positive direction, and get larger and larger. We can extend the system by taking negative numbers to the left of zero, and they get smaller and smaller.

negative direction

−4 −3 −2 −1 0 1 2 3 4

Practical applications of this might well be with a vertical number line

1 representing a thermometer, where 0 would represent 0° Celsius for the freezing point of water. For temperatures above freezing point we would use the positive direction, for temperatures below freezing point we would use the negative direction.
2 representing floors visited by a lift, where 0 would represent the ground floor. For floors above ground level we would use the positive direction, for floors below ground level we would use the negative direction.
3 representing the heights of tide levels at various time intervals, the positive direction being for heights above the given level, the negative direction being for heights below the given level.

Adding directed numbers

Consider the next four examples carefully.

Example 1: $(+4) + (+2) = +6$

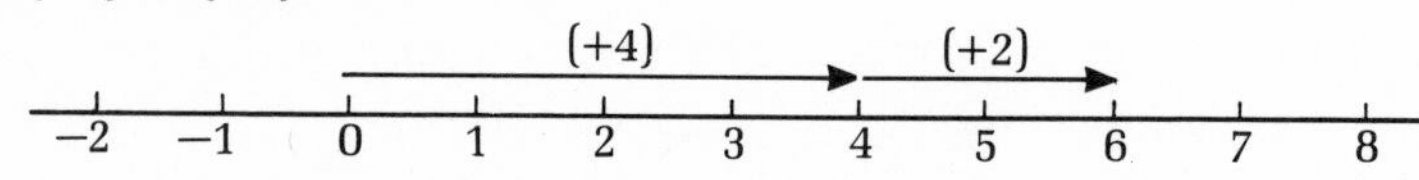

Example 2: $(-4) + (+3) = -1$

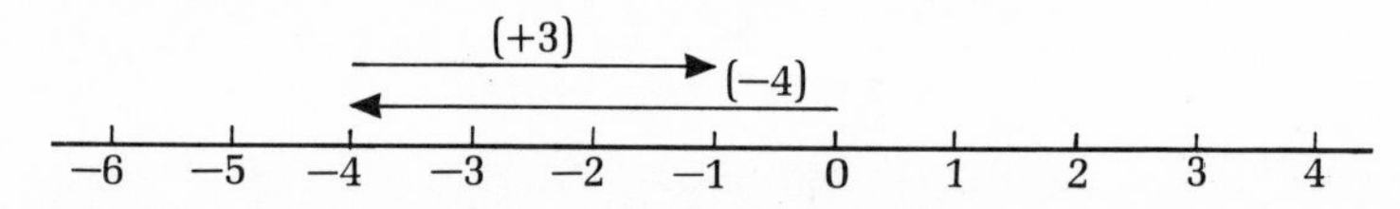

Example 3: (+5) + (−2) = +3

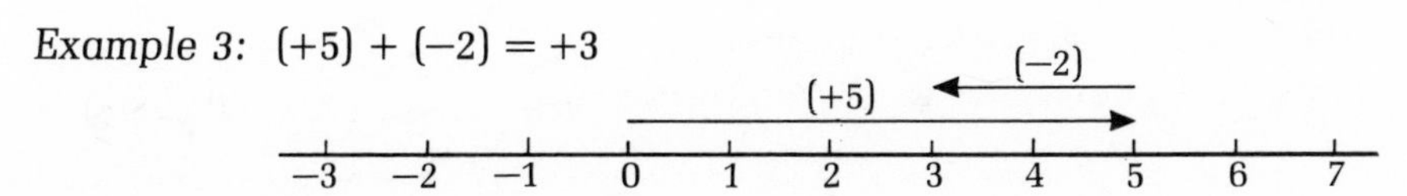

Example 4: (−1) + (−4) = −5

(−4) (−1)

−8 −7 −6 −5 −4 −3 −2 −1 0 1 2

NOTE: If the signs of the two numbers are *alike*, simply add the numbers, irrespective of their signs. The sum then has the same sign as each of the numbers.

Thus in (*1*) 4 + 2 = 6 in (*4*) 1 + 4 = 5, ∴ (−1) + (−4) = −5

BUT: If the signs of the two numbers are *different*, find the numerical difference of the two numbers and the sign of the answer is that of the numerically larger number.

Thus in (*2*) numerical difference of 4 and 3 is 1; since sign of −4 is −, then answer is −1: in (*3*) numerical difference of 5 and 2 is 3; since sign of 5 is +, the answer is +3.

Subtracting directed numbers

Consider the following four examples.

Example 1: (+8) − (+2) = 6 Reverse direction of (+2)

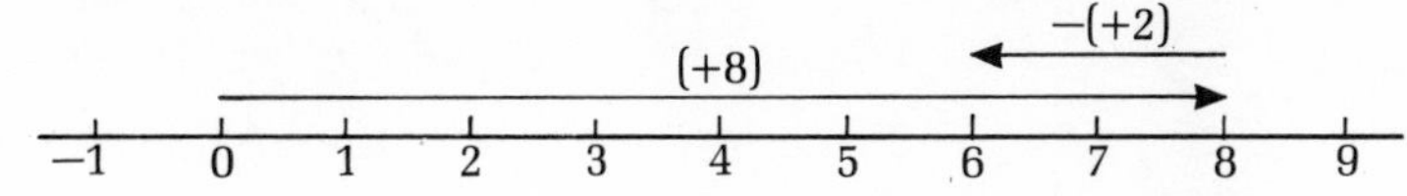

Example 2: (+2) − (+8) = −6 Reverse direction of (+8)

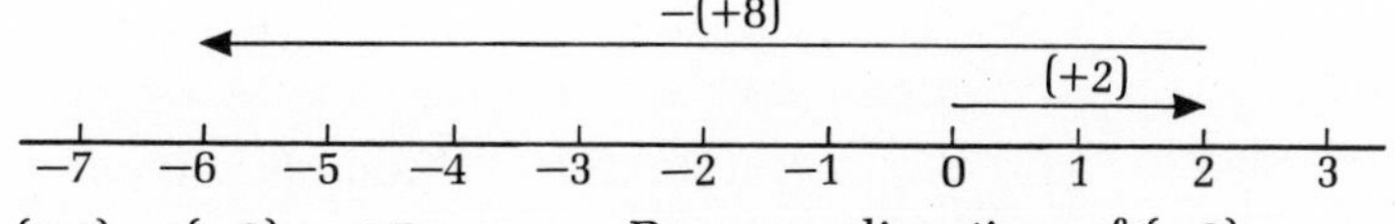

Example 3: (+4) − (−3) = +7 Reverse direction of (−3)

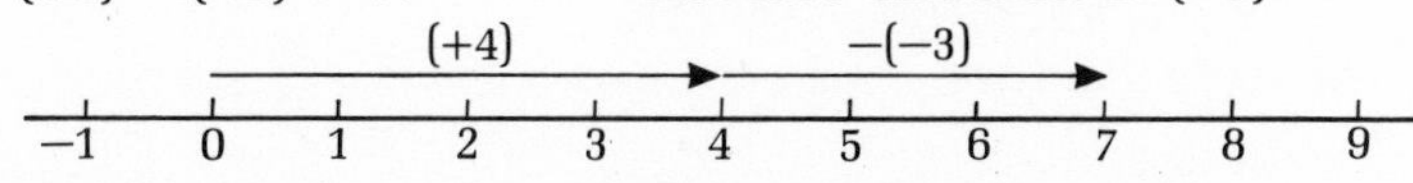

Example 4: (−3) − (−4) = +1 Reverse direction of (−4)

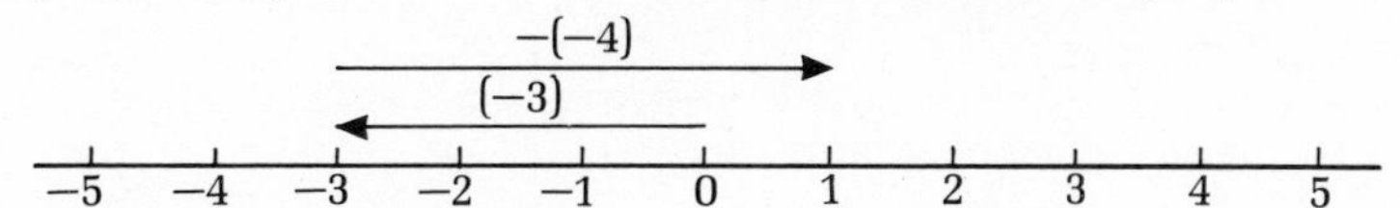

To subtract any directed number from any other number, reverse the sign and add.

(+8) − (+2) = 8 − 2 = 6
(+2) − (+8) = 2 − 8 = −6
(+4) − (−3) = 4 + 3 = 7
(−3) − (−4) = −3 + 4 =1

Multiplication of directed numbers

Make sure you understand the following.

$$3 + 3 + 3 + 3 = 12$$
$$\text{or } 4 \times 3 = 12$$

i.e. the product of two positive numbers is a positive number.

$$(-3) + (-3) + (-3) + (-3) = -12$$
$$4 \times (-3) = -12$$

i.e. the product of a positive number and a negative number is a negative number.

$$(-4) \times (-3) = -(+4) \times (-3)$$
$$= -(-12) = 12$$

i.e. the product of two negative numbers is a positive number.

Division of directed numbers

The rules are like those for multiplication: the division of numbers with like signs yields a positive number whilst division of numbers with unlike signs yields a negative number.

For instance, $\frac{30}{5} = 6$ and $\frac{-30}{-5} = 6$,

but $\frac{-30}{5} = -6$ and $\frac{30}{-5} = -6$.

Questions might well involve practical situations.

Example 1: The temperature is $-6°C$ in New York and is $8°C$ in Manchester. What is the temperature difference between Manchester and New York?

Difference is $8° - (-6°) = 8° + 6° = 14°C$

Example 2: In a harbour at a particular time the depth of the water is 8 m. Later it falls 3 m when the tide goes out. How deep is the water at this time?

$8 - 3 = 5$ m

Example 3: In a large building, the lift is at the ground floor. It goes up 10 floors and descends 12 floors. Where is the lift then in relation to the ground floor?

If we measure upwards as being positive, downwards as being negative, then

$(+10) + (-12) = 10 - 12 = -2$

i.e. the lift is 2 floors below the ground floor (the negative sign tells us that the lift is below ground floor).

If the lift descends a further 3 floors where will it be?

$(-2) + (-3) = -2 - 3 = -5$

i.e. the lift will be 5 floors below the ground floor.

4 Ratio and Proportion

Foundation Level
Ratio, direct proportion, inverse proportion, linear scales.

Foundation Level

Ratio

A ratio is a comparison between two or more similar quantities, where the quantities have the *same* units (*i.e.* all pence, all kilograms, all centimetres, etc.) When we consider the ratio of the quantities then the ratio can be regarded as a fraction:

$\frac{3}{4}$ is written as a ratio 3:4 $\qquad$ $\frac{5}{8}$ is written as a ratio 5:8

Example 1: Two girls, Pam and Jean, receive 90p and £1·50 pocket money respectively. What is the ratio of Pam's money to Jean's money?

Ratio is 90p:150p (units must be the same)

i.e. 90:150

or 3:5 since $\frac{90}{150} = \frac{3}{5}$

Note that there are no units in the final ratio since we are comparing like quantities (same units).

Example 2: A man is 1.85 m tall and his son is 1.5 m tall. What is the ratio of their heights?

Man's height: Son's height is 1.85 m:1.5 m

i.e. 185:150

or 37:30

Example 3: The ratio of the length of a table to its width is 5:3. If the width is 135 cm, what is the length of the table?

Length:width is 5:3, *i.e.* $\frac{5}{3}$

Length is $\overset{45}{\cancel{135}} \times \frac{5}{\cancel{3}_1}$ cm = 225 cm

Example 4: In a disco, the ratio of boys to girls is 3:4. If there is a total of 210 boys and girls at the disco, how many are girls?

Boys:Girls is 3:4, *i.e.* a total of 3 + 4 = 7 parts

$\therefore$ fraction of girls of the total attending is $\frac{4}{7}$

$\therefore \frac{4}{\cancel{7}_1} \times \overset{30}{\cancel{210}} = 120$ girls

Direct proportion

Two quantities are said to be in direct proportion if they increase (or decrease) at the same rate. For instance, if apples are advertised at 30p per pound, then 2lb of apples would cost 60p, 3lb of apples would cost 90p – if we double the amount bought, we double the cost; if we treble the amount, we treble the cost.

Likewise if we halve the amount, we halve the cost etc.

Example: A man pays £1·32 for his newspaper for a total of 6 days. How much would he pay

a for 4 days **b** for 10 days?

The problem can be solved either by finding the cost of one newspaper, then for any number of days; or by the use of fractions.

Either Cost of one newspaper $= \frac{£1{\cdot}32}{6} = 22\text{p}$

a Cost for 4 days $= 4 \times 22\text{p} = 88\text{p}$

b Cost for 10 days $= 10 \times 22\text{p} = £2{\cdot}20$

or **a** Cost is $\frac{4}{6} \times £1{\cdot}32 = \frac{\overset{2}{\cancel{4}}}{\underset{3}{\cancel{6}}} \times \overset{44}{\cancel{132}}\text{p} = 88\text{p}$

b Cost is $\frac{10}{6} \times £1{\cdot}32 = \frac{£13{\cdot}20}{6} = £2{\cdot}20$

Inverse proportion

When an increase in one quantity produces a decrease in a second quantity in the same ratio, then the quantities are said to be in inverse proportion.

Example: If 3 men could tile a roof in 12 days, how long would it take 6 men working at the same rate?

Clearly if we double the number of men, the roof should be tiled in half the time, *i.e.* 6 days.

Ratio of increase of men is 6:3, *i.e.* 2:1

∴ time is decreased in ratio 1:2

Time required $= \frac{1}{2} \times 12 = 6$ days

Conversely, when a decrease in one quantity produces an increase in a second quantity in the same ratio, then the quantities are still in inverse proportion.

Example: How long would one man working on his own take to tile the roof?

Ratio of decrease of men 1:3

Time is increased in ratio 3:1

∴ time required $= \frac{3}{1} \times 12 = 36$ days

Linear scales

Whenever a model is made of some object, then the dimensions of the model are in direct proportion to those of the object. For instance, if the length of the model is $\frac{1}{100}$ of the length of the object, then the width and the height of the model are also $\frac{1}{100}$ of the object's dimensions. In this case the ratio of model to object is 1:100 and the ratio of object to model is 100:1. This ratio is called the linear scale or more often simply the scale.

Example: The wing-span of a model aeroplane is on a scale of 1:200 to that of the real aeroplane. Find the actual wing-span of the aeroplane if that of the model is 8 cm.

$$\text{Wing span} = 8 \times \frac{200}{1} \text{ cm} = 1600 \text{ cm} = 16 \text{ m}$$

A particular use of scales is in the formation of maps. A map uses direct proportion and the scale is usually written in the form of 1:n where n takes any value depending upon what is being represented. To represent, on this page, a map of your town would need a smaller value of n than if you were to represent the whole world.

Every map should have the scale marked on it.

Example 1: On a map, where the scale is 1:20 000, the distance between two towns is 9.5 cm. What is the actual distance between the towns?

$$\begin{aligned}\text{Actual distance} &= 9.5 \times 20\,000 \text{ cm}\\ &= 9.5 \times 200 \text{ m}\\ &= 9.5 \times 0.2 \text{ km}\\ &= 1.9 \text{ km}\end{aligned}$$

Example 2: On a building site, the distance between two bungalows is 460 m. If a drawing uses a scale of 1 cm to 50 m, what is the distance between the bungalows on the map?

$$\text{Map distance} = \frac{460}{50} \text{ cm} = 9.2 \text{ cm}$$

5 Estimation and Approximation

Foundation Level
Estimation.

Intermediate Level
Approximation, decimal places, significant figures.

Foundation Level

Estimation

When you are asked to calculate some multiplication and/or division problems, it is always advisable to estimate an answer by approximating the given numbers to ones which are easy to multiply and/or divide. This is especially the case when dealing with decimals.

Example 1: Calculate 0.28×0.52.
A reasonable estimate is $0.3 \times 0.5 = 0.15$
The calculated result is 0.1456.
You can see that the answer is of the same order as the estimate.

Example 2: Find $82.16 \div 3.16$.
A reasonable estimate is $82 \div 3 = 27\frac{1}{3}$
The calculated result is 26.
Again, the answer is of the same order as the estimate.

Example 3: $\dfrac{0.88 \times 101.2}{40}$
A reasonable estimate is $\dfrac{0.9 \times 100}{40} = \dfrac{90}{40} = 2.25$
The calculated result is 2.2264.
Again, the answer is of the same order as the estimate.

Example 4: How much would 21 postage stamps at 19p each cost?
A reasonable estimate is 20 stamps at 20p each
i.e. $20 \times 20\text{p} = 400\text{p} = £4$.
The calculated cost is $21 \times 19\text{p} = 399\text{p} = £3{\cdot}99$.
The answer is of the right order.

Throughout your experience of mathematics and everyday life you should be able to make reasonable estimates of measures such as lengths, masses, capacities, angles, etc:

How far is it from your home to school? You could measure the distance quite accurately with a trundle wheel or a tape, but you could try to make a good guess such as half a mile, 2 km, or whatever. These would be estimates.

An estimate of the mass of an object could be made by holding the object and comparing it against some known mass, e.g. 1kg of sugar, 1lb of apples, etc.

How long will it take you to travel to Blackpool by car? Clearly it depends upon where you live but you could probably estimate to a half-hour's accuracy.

What is the length of the following line?

If you estimate 5 cm, that would be very good. It is actually 4.8 cm, but you cannot estimate so accurately – or can you?

A good estimate of the following angle would be 45°.

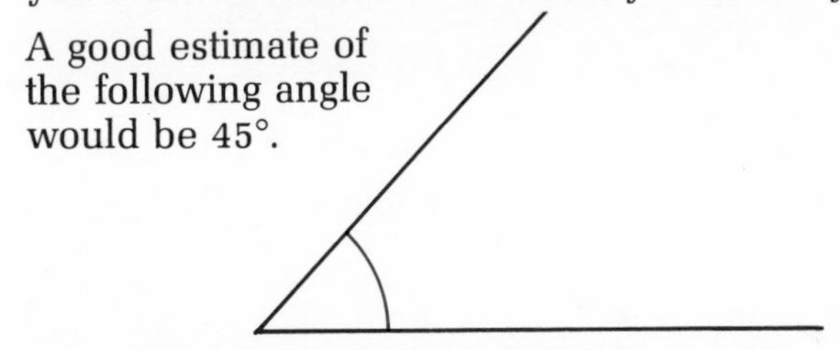

Try to find other estimates, such as your height, mass, reach, etc. and check against your own actual measures.

Intermediate Level

Approximation

In everyday life, a number attached to some unit of measure might well have to be approximated in order to have some realistic meaning.

Decimal places

Example: If a roll of felt is 9 m in length and costs £31·22, how much would a piece 2 m in length cost?

1 m costs $£\frac{31{\cdot}22}{9}$ 2 m costs $£\frac{2 \times 31{\cdot}22}{9} = £6{\cdot}93777...$

Obviously, to have some meaning we would have to say £6·93 or £6·94 since we can only buy something where our lowest currency unit is 1p, *i.e.* £0·01.

In this case we would say £6·94, since £6·937 . . . is nearer to £6·94 than £6·93. Our answer of £6·94 is said to be correct to 2 decimal places of accuracy.

In general, if any number is being corrected to one decimal place, look at the digit in the second decimal place. If this digit is 0, 1, 2, 3 or 4, leave the first decimal place as it is; if the digit in the second place is 5, 6, 7, 8, or 9, increase the digit in the first place by adding 1.

Examples:
25.67 = 25.7 (to 1 d.p.)
8.906 = 8.9 (to 1 d.p.)
115.98 = 116.0 (to 1 d.p.)

Similarly, if you wish to correct any number to 2 decimal place accuracy, look at the third decimal place digit and use the above rule to either leave or increase by 1 the digit in the second decimal place.

Examples:
1.9037 = 1.90 (to 2 d.p.)
0.8762 = 0.88 (to 2 d.p.)
101.091 = 101.09 (to 2 d.p.)
101.098 = 101.10 (to 2 d.p.)

The same procedure is used for any number of decimal place accuracy.

Examples:
13.5682 = 13.568 (to 3 d.p.)
0.08765 = 0.088 (to 3 d.p.)
0.8247 = 0.825 (to 3 d.p.)

Significant figures

Sometimes a number has too many digits to be practical and useful in everyday life. This can be overcome by expressing the number to a particular number of significant figures. For instance, the crowd at a football match might be 14 284. We would still have a very good approximation to the size of the crowd by writing 14 300. This is correct to three significant figures.

We therefore write 14 284 = 14 300 (to 3 s.f.)

If you require an answer to 3 significant figures, look at the fourth digit. If this digit is 0, 1, 2, 3 or 4, the third digit stays as it is; if this digit is 5, 6, 7, 8 or 9, add 1 to the third digit (as above).

Similarly, 14 284 = 14 000 (to 2 s.f.)

NOTE: All remaining digits are replaced by zeros in order to keep the number of the same order as the original. It would be stupid to write 14 284 = 143 or 14.

The *first* significant figure is always a non-zero digit.

Check that you can follow these examples:

89 627 = 89 630 (to 4 s.f.)
21.6293 = 21.63 (to 4 s.f.)
89 627 = 89 600 (to 3 s.f.)
21.6293 = 21.6 (to 3 s.f.)
89 627 = 90 000 (to 2 s.f.)
21.6293 = 22 (to 2 s.f.)
18.068 = 18.07 (to 4 s.f.)
18.068 = 18.1 (to 3 s.f.)
5.009 = 5.01 (to 3 s.f.)
5.009 = 5.0 (to 2 s.f.)

0.00031728 = 0.000317 (to 3 s.f.)
0.00031728 = 0.00032 (to 2 s.f.)
0.00031728 = 0.0003 (to 1 s.f.)
8.00031728 = 8.00 (to 3 s.f.)
8.00031728 = 8.0 (to 2 s.f.)

6 Units of Measurement

Foundation Level
Metric and imperial measures, mass, length, area, volume, capacity, time, common rates of measure, money: use of foreign currency, money: personal and household finance (hire-purchase, taxation, discounts, loans, wages, salaries, VAT or Value Added Tax, simple interest, reading meters and dials, telephone bills).

Higher Level
Compound interest.

Foundation Level

Metric and imperial measures

The metric system is basically a decimal system for common measures and is used throughout the world. The fundamental units for length, mass and time are the metre (m), the kilogram (kg) and the second (s) respectively.
Another system still used in Britain is imperial measure. No doubt you are familiar with such terms as miles, pounds and ounces, pints and gallons, etc.
We will look at the two systems in more detail within this module.

Mass

Mass is the quantity of matter a body contains. Weight is a force and is the effect of gravity on a particular mass. Mass is always constant. The weight of an article varies at different places on the Earth's surface. A man at the top of a mountain weighs less than when he is at the bottom of a mine shaft but he still has the same mass.
The standard unit of mass in the metric system is the kilogram. The prefix kilo- means a thousand, *i.e.* 1000 grams is 1 kilogram.

$$\therefore 1 \text{ kg} = 1000 \text{ g}$$

For larger masses, 1 tonne = 1000 kg.

For smaller masses, 1 g = 1000 mg (milligrams). The prefix milli- means one thousandth.

$$\therefore 1 \text{ mg} = \frac{1}{1000} \text{ g}$$

Common imperial measures are:
16 ounces (oz) = 1 pound (lb)
14 pounds = 1 stone
112 pounds = 1 hundredweight
20 hundredweights (cwt) = 1 ton.

A useful approximation is 1 kg $\simeq$ 2.2lb.
Most questions use the metric system, though questions may also be asked on the imperial system.

Length

In the metric system, the basic unit is the metre (m). The same prefixes as for mass are used (*i.e.* kilo-, centi-, milli-, etc.).

For larger lengths, 1 km = 1000 m.

For smaller lengths, 1 mm = $\frac{1}{1000}$ m or 1000 mm = 1 m.

For some lengths it is more convenient to use centimetres (cm) where

100 cm = 1 m and 10 mm = 1 cm.

In the imperial system the common measures are:

12 inches (in) = 1 foot (ft)
3 feet = 1 yard (yd)
22 yards = 1 chain
10 chains = 1 furlong = 220 yds
8 furlongs = 1 mile = 1760 yds.

Some useful approximations are:

1 in $\simeq$ 2.54 cm, 1 m $\simeq$ 39.4 in, 8 km $\simeq$ 5 miles.

Area

The amount of surface that is covered is always measured in square units.

NOTE: $1\text{ cm}^2 = 1\text{ cm} \times 1\text{ cm} = 10\text{ mm} \times 10\text{ mm} = 100\text{ mm}^2$.

Likewise $1\text{ m}^2 = 100\text{ cm} \times 100\text{ cm} = 10\,000\text{ cm}^2$.

Much confusion occurs when converting units of area from one type to another. Be careful here!

Example 1: Convert 6.2 m^2 to cm^2.

$6.2\text{ m}^2 = 6.2 \times 100 \times 100 = 62\,000\text{ cm}^2$

Example 2: Convert 118 mm^2 to cm^2.

$118\text{ mm}^2 = 118 \times \frac{1}{10} \times \frac{1}{10} = 1.18\text{ cm}^2$

A common measure in the metric system for the area of land is the hectare where 1 hectare = $10\,000\text{ m}^2$.

In the imperial system, $144\text{ in}^2 = 1\text{ ft}^2$
$9\text{ ft}^2 = 1\text{ yd}^2$
$4840\text{ yd}^2 = 1$ acre.

Volume

The volume of a solid is how much space it occupies and this is measured in cubic units.

Standard units are: $10\text{ mm} \times 10\text{ mm} \times 10\text{ mm} = 1\text{ cm}^3 = 1000\text{ mm}^3$
$100\text{ cm} \times 100\text{ cm} \times 100\text{ cm} = 1\text{ m}^3 = 1\,000\,000\text{ cm}^3$

In the imperial system: $12\text{ in} \times 12\text{ in} \times 12\text{ in} = 1\text{ ft}^3 = 1728\text{ in}^3$
$3\text{ ft} \times 3\text{ ft} \times 3\text{ ft} = 1\text{ yd}^3 = 27\text{ ft}^3$

Again, confusion often arises when converting from one unit to another. Be very careful here.

$6.1\text{ cm}^3 = 6.1 \times 10 \times 10 \times 10\text{ mm}^3 = 6100\text{ mm}^3$

$5627\text{ mm}^3 = 5627 \times \frac{1}{10} \times \frac{1}{10} \times \frac{1}{10}\text{ cm}^3 = 5.627\text{ cm}^3$

Capacity

This is the amount of space inside a container and in the metric system, the basic unit of capacity is the litre where 1 litre = 1000 cm^3.

For smaller measures, centilitres or millilitres are used,

i.e. 1 litre = 100 cl or 1000 ml.

In the imperial system, capacity is measured in pints, quarts and gallons, where

2 pints = 1 quart

4 quarts or 8 pints = 1 gallon

Time

The time of day is expressed by using a 12-hour or a 24-hour clock. Most clocks and watches use the 12-hour clock and we distinguish between morning and afternoon by a.m. and p.m. (ante meridiem, before midday, and post meridiem, after midday, respectively).

Timetables are often expressed using the 24-hour clock. A 24-hour clock starts at midnight (0000 hours) and goes to 2400 hours, i.e. representing one complete day.

5.30 a.m. is 0530 hours in the 24-hour clock system

7.45 p.m. is 1945 hours in the 24-hour clock system

Conversely 2052 hours is 8.52 p.m.

Common rates of measure

Units are combined to form common rates of measure. The most important is that of average speed:

$$\text{Average speed} = \frac{\text{Total distance travelled}}{\text{Total time taken}}$$

If the distance is in kilometres (km) and time in hours (h), then average speed is in kilometres per hour and written as

km per hour, or km/h, or km h^{-1}.

In questions on average speed, always ensure that the units of distance and time are converted to those which are required.

Example: If a car covers 800 m in 1 min, what is the average speed in km/h?

800 m = 0.8 km

$$1 \text{ min} = \frac{1}{60} \text{ h}$$

$$\therefore \text{average speed} = \frac{0.8}{\frac{1}{60}} \text{ km/h} = 0.8 \times 60 \text{ km/h} = 48 \text{ km/h}$$

Another common rate of measure is that of density:

$$\text{Density} = \frac{\text{Mass}}{\text{Volume}}. \quad \text{(units could be g/cm}^3\text{, etc.)}$$

Money: use of foreign currency

When you visit another country on holiday, you need to change pounds sterling into the currency of that country. The rate of exchange varies day by day. Banks and daily newspapers show these rates and you usually have to take a chance on when it is best to get the most favourable rate.

Questions normally involve a table of information or reading from a graph.

Example 1: **a** If on a particular day £1 = \$1.75 (\$ is the symbol for an American dollar), how many dollars would you get for £80?

You would get 1.75×80 dollars, i.e. \$140.

b If an article in the USA costs \$36, what is it worth in pounds sterling (to nearest penny)?

Cost of article = \$36 = $\$\frac{36}{1.75}$ = £20·57 (to nearest penny)

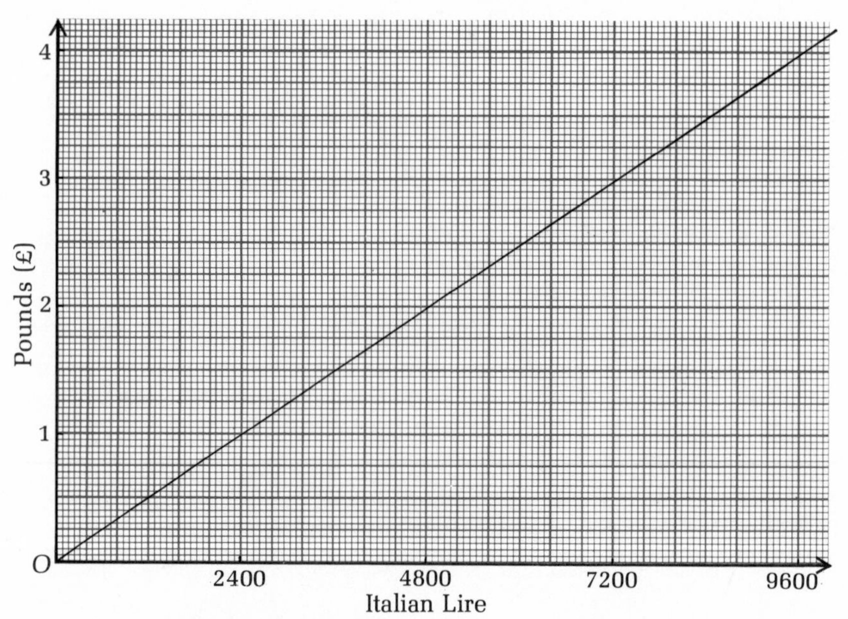

Example 2: The graph shows the rate of exchange for Italian Lire against the pound.

a What is the value of £2·50 in lire?
From graph, 6000 lire.

b How many lire would you get for £90?
From graph, £3 is equivalent to 7200 lire.
∴ £90 is equivalent to $30 \times 7200 = 216\,000$ lire

Money: personal and household finance

Questions could be set on hire-purchase, taxation, discounts, loans, wages, salaries, VAT (Value Added Tax), simple interest, reading meters and dials. The following are likely situations.

Hire-purchase

A car is advertised at £4995 for cash. On hire-purchase a deposit of £99 is required plus £40·80 weekly for 3 years. How much more is paid by taking out hire purchase rather than paying cash? (Assume 52 weeks per year.)

$$\text{HP is } £(99 + 3 \times 52 \times 40.80) = £(99 + 6364.80) = £6463{\cdot}80$$

$$\therefore \text{ excess paid} = £6463{\cdot}80 - £4995 = £1468{\cdot}80$$

Taxation

A man earns £15 625 per year. He has allowances of £3905 that are not taxable. He pays tax at the standard rate of 25% on taxable income. Calculate the income tax he pays in a year.

$$\text{Taxable income} = £(15\,625 - 3905) = £11\,720$$

$$\therefore \text{ tax paid} = £11\,720 \times \frac{25}{100} = £2930$$

Discounts

a A twin-deck portable hi-fi was £90, but is on sale at a discount price of 40% off. What will it now cost?

$$40\% \text{ of } £90 = £\frac{40}{100} \times 90 = £36$$

$$\therefore \text{ discount price is } £(90 - 36) = £54$$

b An entertainment centre was offered at £120 in a sale, after a 25% discount on the original selling price. What was this normal selling price?

£120 is 75% of the original selling price.

$$\therefore \text{ 1\% of original selling price} = £\frac{120}{75}$$

$$\therefore \text{ original selling price (100\%) is } £100 \times \frac{120}{75} = £160$$

Loans

A woman borrows £250 from a bank that charges interest of 16% per year for the loan. She agrees to repay the loan over a year by equal monthly repayments. How much is each repayment? (Answer to nearest penny.)

$$\text{Interest for one year} = £250 \times \frac{16}{100} = £40$$

$$\text{each repayment} = £\frac{290}{12} = £24{\cdot}17$$

Wages

A canteen assistant receives £2·18 per hour for a $37\frac{1}{2}$-hour week. She is paid overtime at time-and-a-half. What is her

a basic weekly wage?

b wage for a 42-hour week?

c If she earned £101·37 in a week, how many hours overtime had she worked that week?

a Basic wage is £2·18 × 37.5 = £81·75

b Wage is £81·75 + £2·18 × 4.5 × 1.5 = £81·75 + £14·715 = £96·465 = £96·47 (to nearest penny)

c Overtime = £101·37 − 81.75 = £19·62

Overtime per hour = £2·18 × 1.5 = £3·27

No. of hours overtime $= \frac{£19·62}{£3·27} = 6$

Salaries

A lecturer earns £11 600 per year. His salary is paid in 12 monthly instalments.

a How much does he earn per month?

b If, in fact, he works 30 hours per week, find his hourly rate, assuming 52 weeks in a year.
Give each answer to the nearest penny.

a Monthly pay $= £\frac{11\,600}{12} = £966·67$ (to nearest penny)

b Hourly pay $= £\frac{11\,600}{52 \times 30} = £7·44$ (to nearest penny)

VAT (Value Added Tax)

This is a tax added to some goods and some labour costs and is determined by the government. At present the rate of VAT is 15%.

a A plumber's bill was £69·20 for the goods and labour, but 15% VAT had to be added. What was the total bill?

VAT $= £69·20 \times \frac{15}{100} = £10·38$

Total bill was £(69·20 + 10.38) = £79.58

b Another bill was £50·60, which included the VAT. What was the cost before VAT was added?

£50·60 is equivalent to 115% of cost

$\therefore$ 1% is $£\frac{50.6}{115}$

$\therefore$ 100% is $£\frac{50.6}{115} \times 100$, *i.e.* cost before VAT = £44

Simple interest

Simple interest is given by the formula $I = \frac{PRT}{100}$ where I is the interest, P is the Principal (the amount borrowed), R is the rate % per annum, T is time in years. The interest is the same each year.

a Calculate I when £300 is borrowed @ 12% for 2 years.

$$I = £\frac{300 \times 12 \times 2}{100} = £72$$

b If the interest on £250 is £60 for 3 years, what was the rate?

$$60 = \frac{250 \times R \times 3}{100}$$

Multiply both sides of the equation by 100:

$$60 \times 100 = \frac{250 \times R \times 3}{\cancel{100}_1} \times \cancel{100}^1$$

Now divide both sides by 250 × 3:

$$\frac{60 \times 100}{250 \times 3} = \frac{250 \times R \times 3}{250 \times 3}$$

$$\therefore \quad \frac{{}^4\cancel{20}\cancel{60} \times \cancel{100}^2}{{}_1\cancel{5}\cancel{250} \times \cancel{3}_1}\% = R$$

$\therefore 8\% = R \qquad$ or $\qquad R = 8\%$

Reading meters and dials

The gas and electricity supplies to households use meters to measure the amount used. It is proposed that water too will eventually be measured by meter. There are digital meters and there are dial meters. These are usually read quarterly (every three months), and the householder pays for what he/she has used.

Example: A gas meter registers → | 8 | 0 | 1 | 4 | 100s of cubic ft

The previous reading was → | 7 | 6 | 5 | 7 | 100s of cubic ft

∴ 8014 − 7657 = 357 (100s of cubic feet) had been used.

Each of these units is 1.026 therms and the cost per therm is 38 pence. British Gas also add a standing charge of £8·20 per quarter, independent of how much gas has been used. Calculate the bill for a quarter.

Cost = 357 × 1.026 × 38 pence	= £139·19
Standing charge	= £ 8·20
∴ Total bill	£147·39

Telephone bills

Telephone calls are metered and units are currently charged at 4·40p per unit. There is also a rental fee per quarter. The total of all these charges is subject to VAT.

Example: In a particular quarter, the rental fee is £17·25 and 499 units have been metered. Find the total payable to the nearest penny.

499 units @ 4·40p = £21·956

Rental = £17·25

Total (before VAT) = £39·206

VAT $\frac{15}{100} \times$ £39·206 = £5·8809

Total payable = £45·09 (to nearest penny)

Higher Level

Compound interest

In simple interest, the interest is always the same. In compound interest, the interest for a year is added to the principal before calculating the next year's interest. Therefore the interest is always increasing if the rate remains constant.

Example: Calculate the total interest after 3 years when £400 is invested at 9% per annum.

		£	
Principal for 1st year	=	400	
Interest for 1st year	=	36	$\left[\frac{9}{100} \times 400\right]$
Principal for 2nd year	=	436	
Interest for 2nd year	=	39.24	$\left[\frac{9}{100} \times 436\right]$
Principal for 3rd year	=	475.24	
Interest for 3rd year	=	42.7716 ..	$\left[\frac{9}{100} \times 475.24\right]$
Amount after 3 years	=	518.0116	
Original principal	=	400	
Interest	=	118.0116	

∴ Interest = £118·01 (to nearest penny)

There are tables from which this answer could have been read off much more quickly, but the above method or the use of the formula (see below) are more likely to be required in an examination.

If A is the total amount of money after n years at r% per annum on £P (principal) then

$$A = P\left(1 + \frac{r}{100}\right)^n$$

Example: Calculate the total interest after 3 years when £400 is invested @ 9% per annum.

$$A = £400 \times \left(1 + \frac{9}{100}\right)^3 = £400 \times 1.295029 = £518{\cdot}0116$$

$= £518{\cdot}01$

∴ Interest = £(518·01 − 400) = £118·01

7 Simple Solid Figures

Foundation Level

Descriptions of a cube, cuboid, prism, cylinder, pyramid, triangular pyramid, cone, sphere.

Foundation Level

Cube

A cube is a solid shape with six equal square sides. It is represented in two dimensions as shown. Dotted lines indicate edges that cannot be seen.

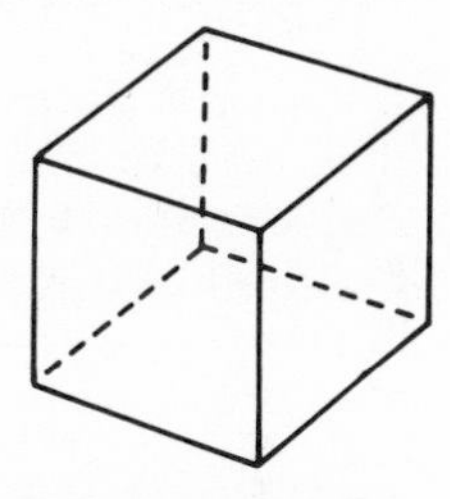

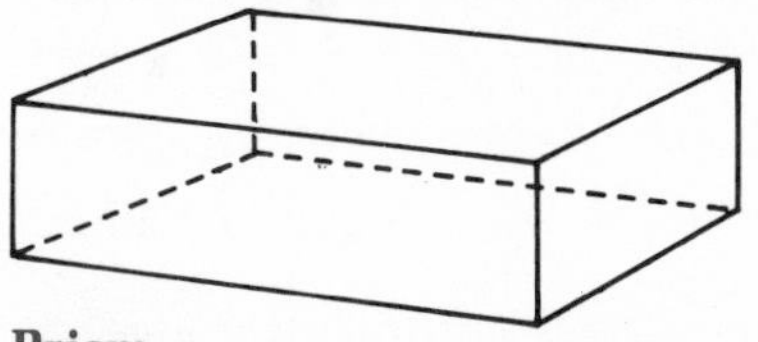

Cuboid

A cuboid is a solid shape with six rectangular sides in which only opposite faces are equal.

Prism

A prism is any solid shape whose ends are equal and parallel and has a constant cross-section. The cuboid shown above is a rectangular prism. A triangular prism is shown alongside.

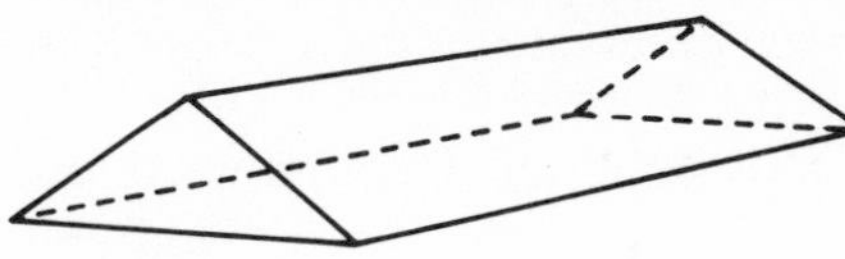

Cylinder

A cylinder is a solid whose uniform cross-section is always a circle.

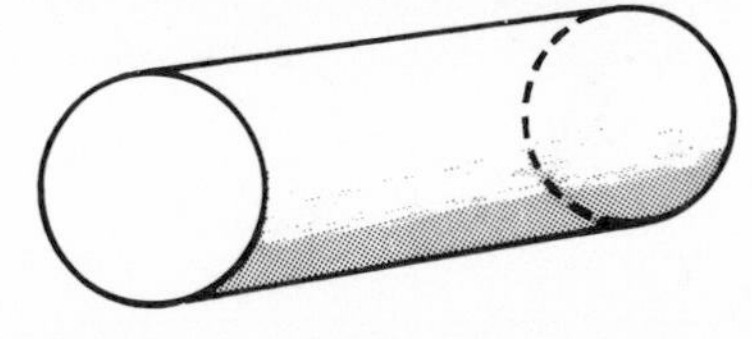

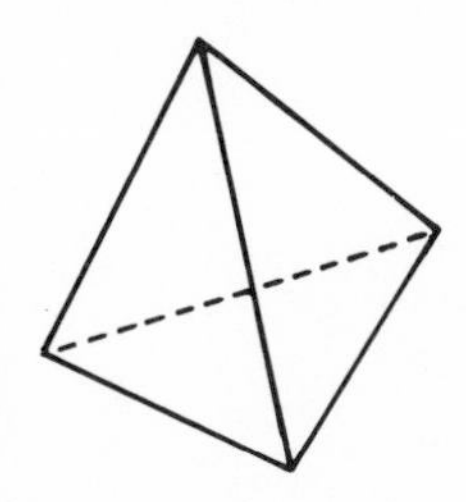

Pyramid

A pyramid is a solid shape on any polygonal base with sloping sides meeting at an apex (a point). The most common pyramids are those with square, triangular* or rectangular bases. When a pyramid has its apex above the centre of the base, it is said to be a right pyramid.

* also called a tetrahedron.

Cone

A cone is a solid shape which tapers to a point from a circular base.

If the cone is said to be a *right circular* cone, then the apex is vertically above the centre of the base.

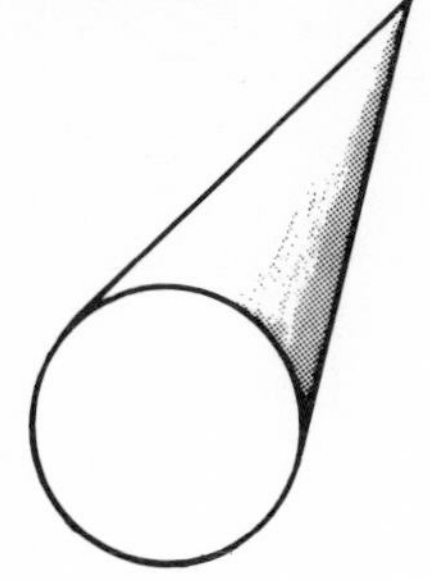

Sphere

A sphere is a round solid body such as a ball.

8 Symmetry

Foundation Level
Symmetric properties, nets of a cube, cuboid, prism.

Foundation Level

Symmetric properties

Imagine you have a child's post-box into which solids in the forms of cubes, cuboids and prisms can be pushed through certain holes.

If you had a square opening through which the cube could pass, how many ways could this be done? Due to the symmetry of the cube, this could be done in 24 ways. Think of the cube as though it were a dice. The upper face could show any of the six faces and each face could be rotated in four different ways.

Suppose there was a rectangular opening through which a cuboid could pass. In how many ways could this be done? In this case, the answer would depend on the dimensions of the cuboid. Suppose the cuboid is 10 cm by 6 cm by 2.5 cm and that the opening is 10.1 cm by 2.6 cm. The cuboid could be pushed into the box in only four different ways, since by symmetry each end face that measures 10 cm by 2.5 cm could be rotated twice giving the four ways.

For any prism, the number of ways depends upon the cross-section of the prism and the shape of the opening in the post box – you should investigate this for yourself.

Nets: cubes

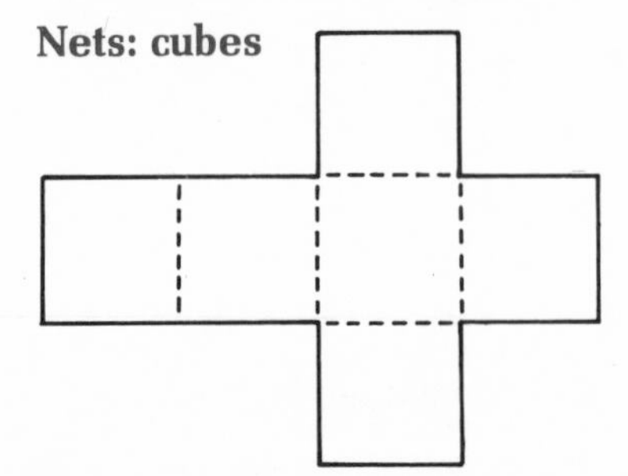

If this shape was drawn on card, cut out, and folds made along the dotted line, then the solid formed would be a cube. The drawing is called a *net* of a cube.

The cube would have six faces, so is it possible to have more than one net for a cube? The answer is, yes. You should try to find how many nets can be drawn in order to form a cube. There are 35 ways in which non-congruent shapes can be drawn by having 6 squares joined together, edge to edge, but not all will give the net of a cube.

It would be impossible to form a cube from this shape:

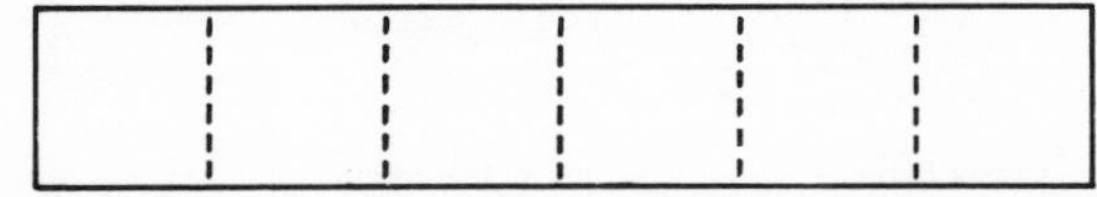

But it would be possible to form a cube from this one:

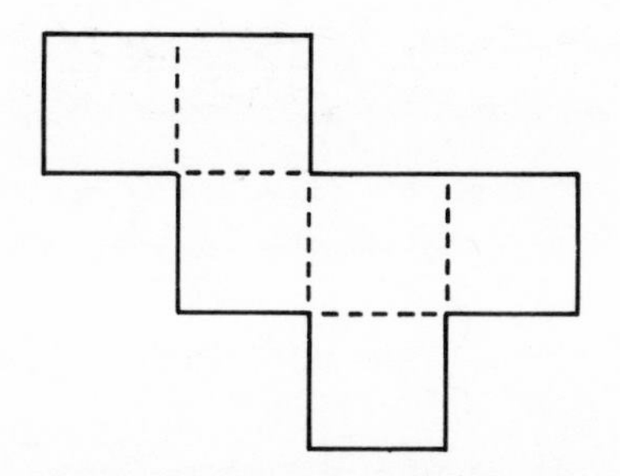

Nets: cuboids

The net of a cuboid is different, since only opposite faces of a cuboid are the same size.

Draw a possible net of a box that measures 10 cm by 6 cm by 2.5 cm.

Here is an example:

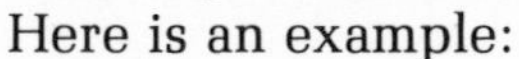

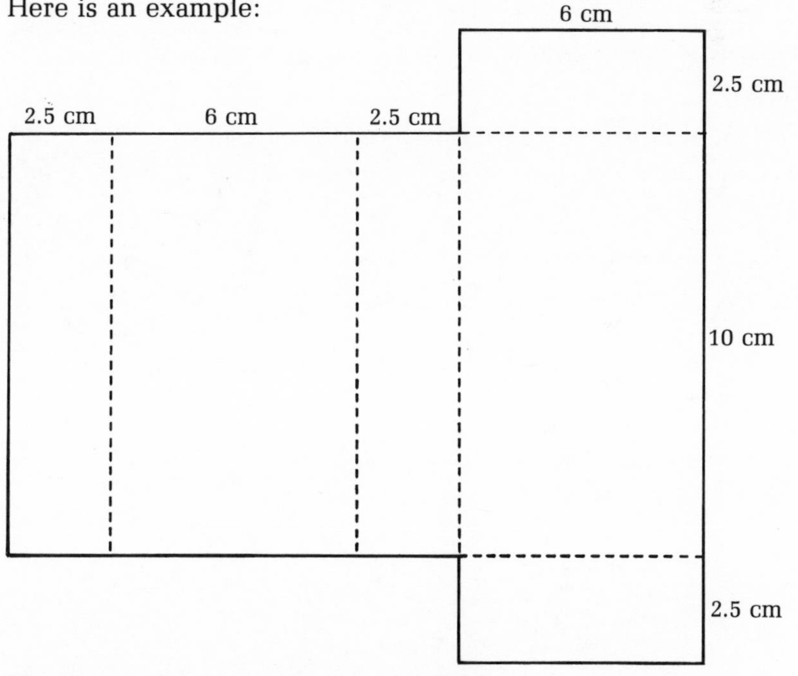

There are, of course, other possibilities.

Nets: prisms

This net forms a prism when folded to form a solid.

Other nets would do so, too.

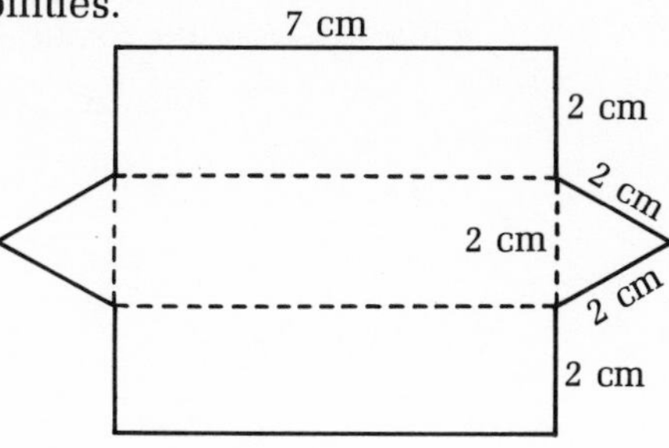

9 Perimeter, Area and Volume

Foundation Level
Perimeter, area, circumference of a circle, volume of a cuboid.

Intermediate Level
Area of parallelogram, trapezium, circle, volume of prism, cylinder.

Higher Level
Curved surface area of cylinder, cone, sphere, volume of pyramid, cone, sphere, arc length and area of a sector of a circle.

Foundation Level

At this level we are concerned only with particular shapes and particular types of measures.

Perimeter

The perimeter is the total distance around the edge of some flat shape.

NOTE: The units for perimeter are those of length and so should always be in mm, cm, m, etc.

Example 1:

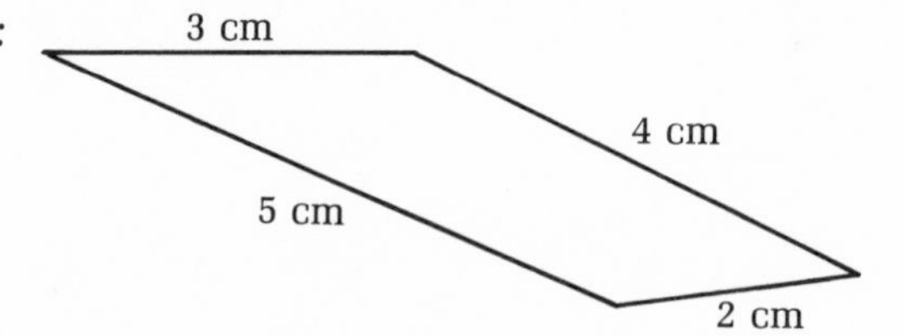

The perimeter is (3 + 4 + 2 + 5)cm = 14 cm

Example 2:

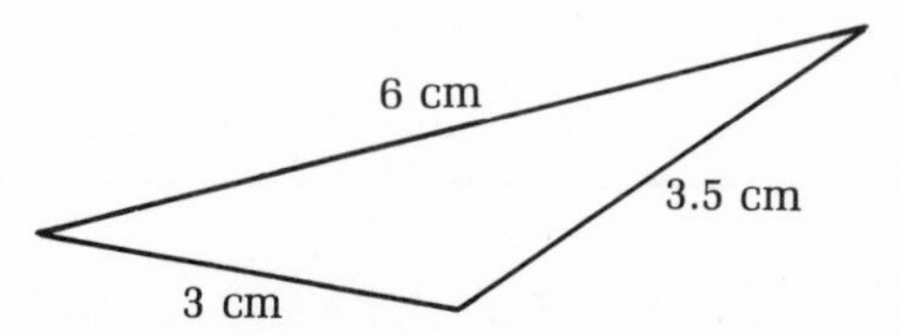

The perimeter of the triangle is the sum of the lengths of the three sides, *i.e.* (3 + 6 + 3.5)cm = 12.5 cm.

Example 3: A rectangle has two pairs of equal sides.

In general:

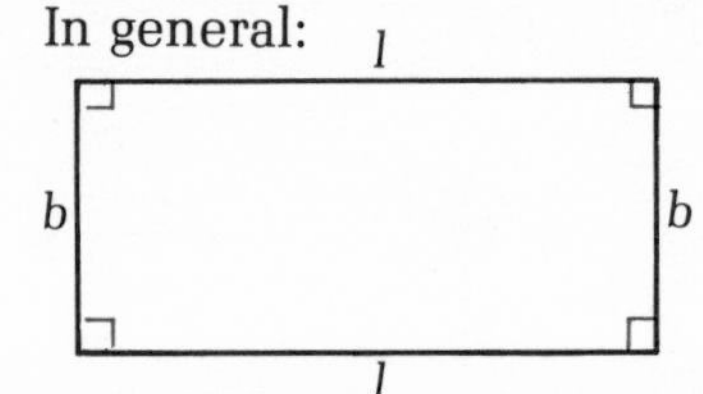

If l is the length and b is the breadth, then

$$\text{perimeter} = l + b + l + b = 2l + 2b$$

Area

The area of any flat shape is the amount of square units that it contains.

The area of the rectangle on page 40 is the product of length and breadth or the product of two adjacent sides, *i.e.*

$$\text{Area of a rectangle} = l \times b.$$

NOTE: The units of area will always be mm^2, cm^2, m^2, etc.

Exercise: If a rectangular piece of paper is 20 cm by 12 cm, its area is $20 \text{ cm} \times 12 \text{ cm} = 240 \text{ cm}^2$.

This is read 240 square centimetres or 240 centimetres squared.

You will be expected to remember the formula for the area of a rectangle.

The following formula should also be learned but will probably be given on a formula sheet for the examination.

The area of a triangle is $\dfrac{\text{length of base} \times \text{vertical height}}{2}$,

i.e. $\text{Area} = \dfrac{b \times h}{2}$

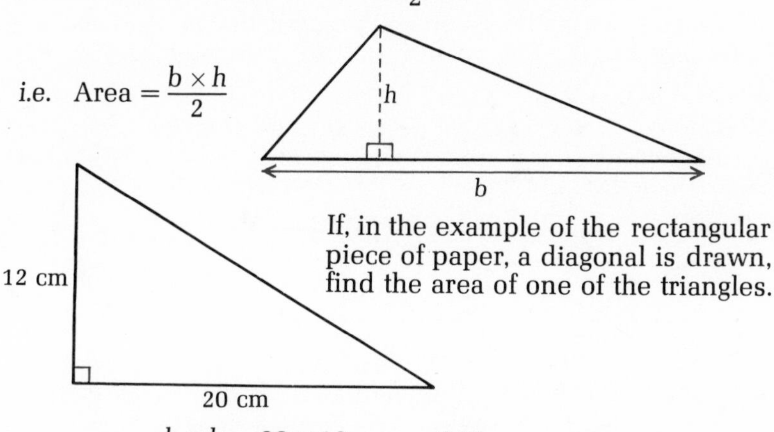

If, in the example of the rectangular piece of paper, a diagonal is drawn, find the area of one of the triangles.

$$\text{Area} = \frac{b \times h}{2} = \frac{20 \times 12}{2} \text{ cm}^2 = \frac{240}{2} \text{ cm}^2 = 120 \text{ cm}^2$$

Clearly the piece of paper was cut into two equal portions.

A second example could be to find the area of the following triangle.

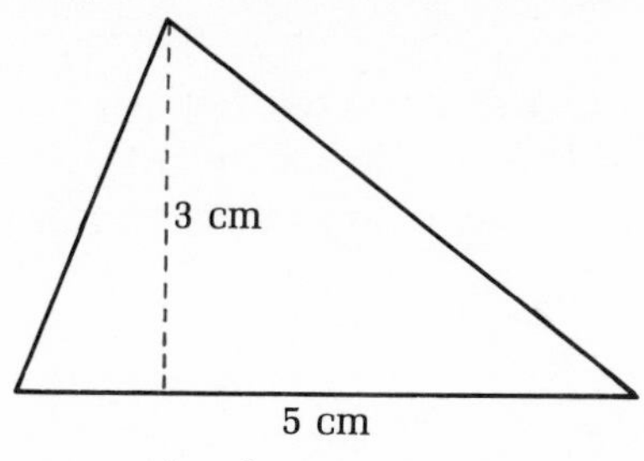

$$\text{Area} = \frac{b \times h}{2} = \frac{5 \times 3}{2} \text{ cm}^2 = \frac{15}{2} \text{ cm}^2 = 7.5 \text{ cm}^2$$

Circumference of a circle

The circumference of a circle is the *perimeter* of the circle.

radius
diameter
centre

The circumference, C, of any circle of diameter d or radius r is given by $C = 2\pi r$ (or πd) where π is a special type of number, called an irrational number. π is approximately 3.142. A useful approximate fraction for π is $\frac{22}{7}$.

Example: Find the circumference of a circle of radius 5 cm, given that π is 3.14.
$C = 2\pi r = 2 \times 3.14 \times 5$ cm $= 31.4$ cm

The circumference will have the same units as the radius.

NOTE: In examinations, you will be given the value of π to be used. Be careful to use the given value and not your own familiar π.

Volume of a cuboid

A cuboid is a rectangular prism, i.e. the uniform cross-section is a rectangle and the volume of any prism is the area of uniform cross-section $\times$ length of prism.

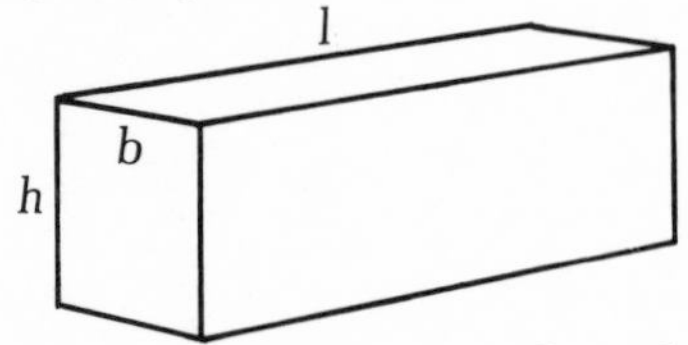

$$\text{Volume} = (b \times h) \times l$$
$$= b \times h \times l$$

The units for volume will be mm^3, cm^3 or m^3, *i.e.* cubic units.

Example: Find the volume of a cuboid 5 cm by 8 cm by 6 cm.
Volume $= (5 \times 8 \times 6)\ cm^3$
$= 240\ cm^3$ or 240 cubic centimetres.

You will need to remember how to find the volume of a cuboid.

Intermediate Level

In the examination, formulae will be given for all of the following, though clearly it is better for you if you know them already.

Area of a parallelogram

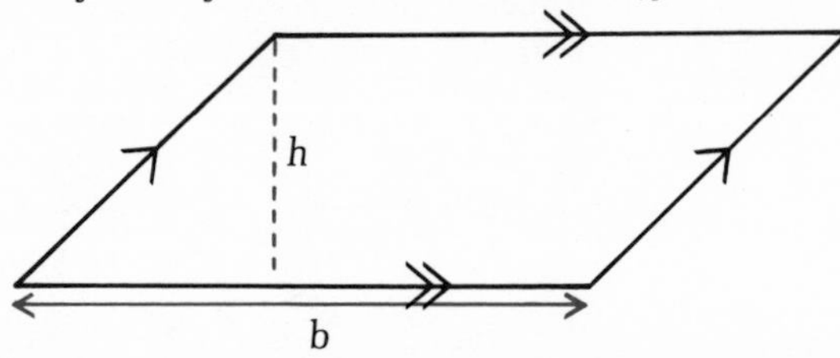

If b is the base length and h the vertical (or perpendicular) height,

Area of a parallelogram $= b \times h$.

If $b = 10$ cm, $h = 6$ cm, area of parallelogram $= 10 \times 6 = 60$ cm^2

Area of a trapezium

A trapezium is a quadrilateral with one pair of opposite sides parallel.

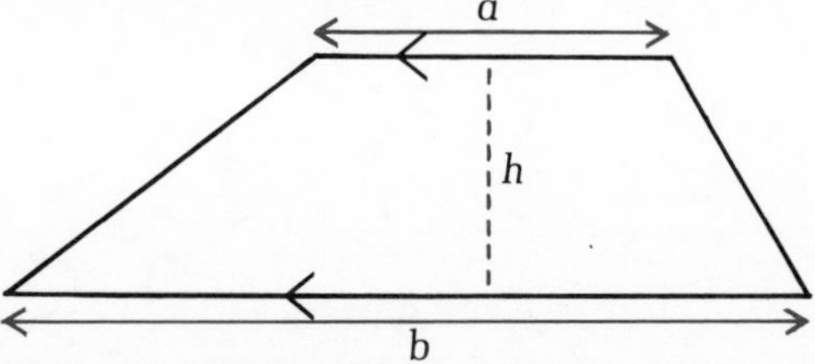

The area of a trapezium is half the sum of the parallel sides $\times$ perpendicular distance between the parallel sides, *i.e.*

Area of trapezium $= \frac{1}{2}(a + b) \times h$

Example 1: Find the area if $a = 6$ cm, $b = 10$ cm and $h = 4$ cm.

Area of trapezium $= \frac{1}{2}(6 + 10) \times 4$

$= \frac{1}{2} \times 16 \times 4 = 32$ cm^2

Example 2: If the area of a trapezium is 70 cm^2, $a = 18$ cm and $h = 5$ cm, find b.

$\frac{1}{2}(18 + b) \times 5 = 70$

$\therefore (18 + b) = \frac{70 \times 2}{5} = 28$

$\therefore b = 28 - 18 = 10$ cm

Area of a circle

This is given by Area $= \pi r^2$ where r is the radius and π is approximately 3.142.

Example: Taking π to be 3.14, find the area of a circle of radius 25 cm, giving your answer to 3 significant figures.

Area $= 3.14 \times 25 \times 25$ cm$^2 = 1962.5$ cm$^2 = 1960$ cm^2 (3 s.f.)

In the exam, you will probably find that questions are set using a combination of the various Area formulae covered in this section.

Volume of a prism

The volume of a prism is the area of uniform cross-section $\times$ the length of the prism.

Example: A girder has a uniform cross-section in the form of a trapezium. Find the volume of the girder in m^3.

50 cm
80 cm
2 m
120 cm

$$\text{Area of cross-section} = \tfrac{1}{2}(0.5 + 1.2) \times 0.8 \text{ m}^2 = 0.68 \text{ m}^2$$

$$\therefore \text{ volume of girder} = 0.68 \times 2 \text{ m}^3 = 1.36 \text{ m}^3$$

Volume of a cylinder

The volume of a cylinder is the area of uniform cross-section × length of cylinder. If r is the radius of circular section and l is the length of the cylinder,

$$\text{Volume of cylinder} = \pi r^2 \times l = \pi r^2 l$$

Example: A cylindrical can has a radius of cross-section of 4 cm and is 15 cm long. Find the volume of the can. (Take π to be 3.142). Give your answer to 1 d.p.

$$\text{Volume} = \pi \times 4^2 \times 15 = 3.142 \times 16 \times 15 \text{ cm}^3 = 754.08 \text{ cm}^3 = 754.1 \text{ cm}^3 \text{ (to 1 d.p.)}$$

Higher Level

Formulae will normally be given in the examination. You must be able to apply the formulae for the following surface areas: cuboid, cylinder, prism, pyramid, cone and sphere.

Cuboids, prisms and pyramids have basically been covered.

Curved surface area of a cylinder

Imagine an open paper cylinder of radius r and length l being cut along its length and then opened out.

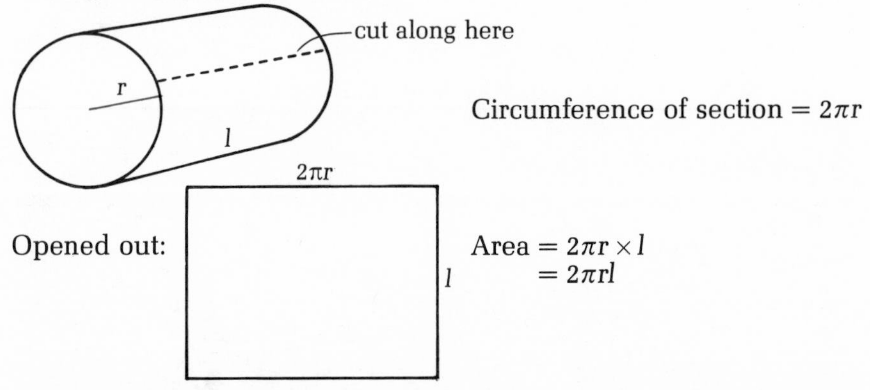

NOTE: Remember, this is the area of the *curved* surface. In some questions, you might need to add on one or both ends of an open or closed cylinder if you are considering the total surface area of an object.

Curved surface area of a cone

If l is the *slant* length and r the base radius, then the *curved* surface area of a cone is πrl.

Be careful here: you are concerned with the slant length and *not* the vertical height.

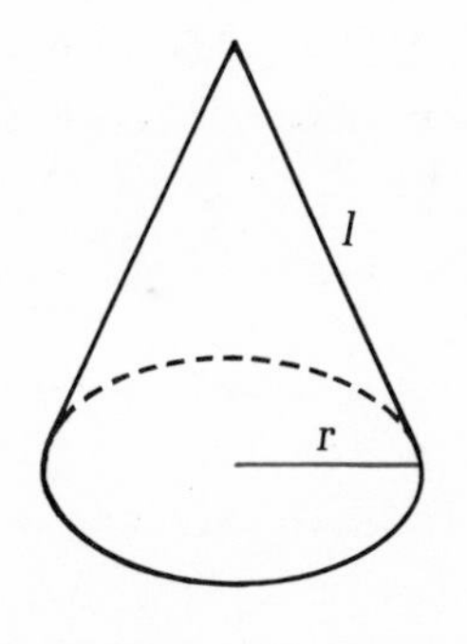

Curved surface area of a sphere

If r is the radius of the sphere, then the curved surface area of the sphere is $4\pi r^2$.

Volume of a pyramid

The volume of a pyramid is given by $\frac{1}{3}$ × Area of base × vertical height of the pyramid.

$$V = \frac{1}{3} Ah$$

NOTE: A pyramid could be on any base – a triangle, or a rectangle, or a hexagon, etc. The base need not be a square. Read questions in the examination very carefully.

Volume of a cone

This is just a special case of the previous example, so its volume is $\frac{1}{3}$ × Area of base × vertical height of the cone.

$$V = \frac{1}{3}\pi r^2 h$$

where r is the radius of circular base, h the vertical height.

Volume of a sphere

If r is the radius of the sphere, then the volume of the sphere is $\frac{4}{3}\pi r^3$.

In the examination, questions will be set on practical combinations of any of the above solids.

Arc length and area of a sector of a circle

If s is the length of the arc which subtends an angle of $\theta°$ at the centre of a circle of which the arc is a part, then $s = \frac{\theta}{360} \times 2\pi r$.

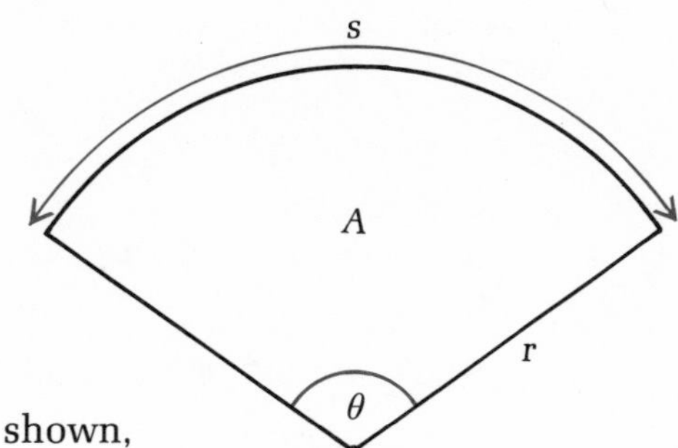

If A is the area of the sector as shown,

$$A = \frac{\theta}{360} \times \pi r^2.$$

The arc length:circumference of circle is in the ratio of θ:360.

Area of sector:area of circle is in the ratio of θ:360.

10 Statistics and Probability

Foundation Level
Data, tabular form, pictorial representation, bar charts, pie chart, measures of averages (mean, median, mode), probability.

Intermediate Level
Frequency distributions, histograms, simple combined probabilities, probability tree diagrams.

Higher Level
Continuous variables, range, cumulative frequency diagrams, quartiles.

Statistics is the collection, ordering, interpretation and analysing of data.

Foundation Level

Data

Data collected first-hand by questioning people, by carrying out experiments or by observation is known as *primary* data.

Data from other sources (such as books, newspapers or periodicals) or collected by other people is known as *secondary* data.

The quantity being investigated is called the *variable* and a member of the variable is called a *variate*.

The data is normally organised in tabular or pictorial form so that it is easy to make simple inferences about the data.

Tabular form

The simplest way of ordering data is by means of a frequency distribution table. In such a table the quantity being measured is recorded against the number of times that it occurs. For example, if a dice is tossed 20 times and the results recorded, then an example of a frequency distribution table could be as shown. Here the variable is the score of the dice, the 6 variates are the scores 1, 2, 3, 4, 5, 6.

Score of dice	1	2	3	4	5	6
Frequency	5	4	2	3	3	3

A score of 4 appears 3 times.
It is easy to pick out the information.

Pictorial representation

Another way in which information can be presented in a more meaningful manner is by means of a picture. As the saying goes, 'a picture is worth a thousand words'. We shall consider three types, noting also their disadvantages.

Pictogram

Sometimes called a pictograph or ideograph, this uses one small symbol to represent an amount of some quantity that is being observed.

Example: Suppose 120 fifth-formers are questioned about their eating habits at lunch-time and the following results are obtained: 60 have school lunch, 36 have sandwiches, 18 go home, and 6 have no lunch.

Let  represent 12 fifth-formers.

Then the pictogram could be represented by

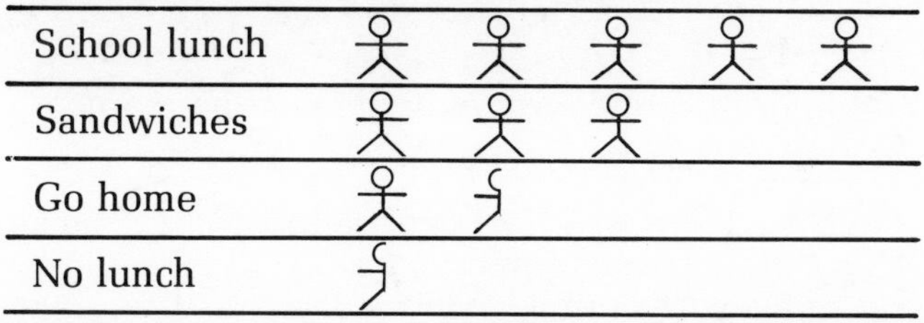

School lunch	
Sandwiches	
Go home	
No lunch	

The information can be presented vertically or horizontally. It is easy to understand, but the main disadvantage is apparent when the frequencies to be depicted are not multiples or easy part-multiples of the given symbols. For instance, it would be much more difficult to represent the above if original data had been 61, 35, 17 and 7 respectively.

Bar chart

Generally regarded as the most popular method of pictorial representation, bar charts are easy to understand and to apply.

The proportionate bar chart

In this, a single bar of constant width is divided into parts, the length of each part being proportionate to the frequency of the items under consideration. The bar may be drawn vertically or horizontally.

The lunch-time information in the above pictogram could be shown as:

60	36	18	6
School lunch	Sandwiches	Go home	No lunch

The diagram is accurate, easy to construct and understand.

It is also often used to represent percentage components to form a bar that is a 100% proportionate bar chart.

Simple bar chart (or bar graph)

This also may be constructed vertically or horizontally. The information is represented by a series of bars, each bar being of equal width. Usually the bars are separated by a space of about a half width of a bar and the length of a bar is proportional to the frequency of the data that it represents.

Using the same lunch-time information as before:

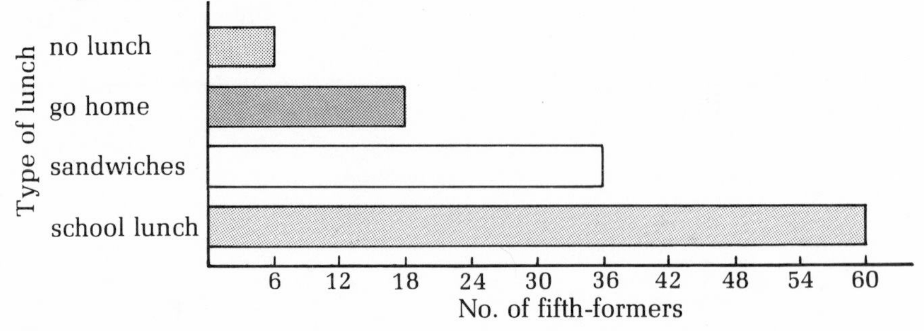

The simple bar chart is easy to understand and apply.

Pie chart

A pie chart (also called a pie-graph) displays the frequencies proportionate to the areas of the sectors of a circle, *i.e.* the sector angles at the centre of the circle.

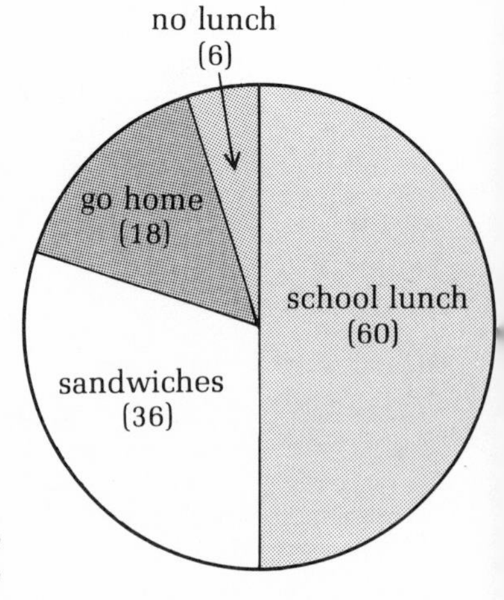

The pie chart has the advantage of being relatively easy to interpret but it is harder to construct than a bar chart. A pie chart is effective when no more than 6 sectors are shown. At this level, the frequencies involved will be factors of 360.

Since the angle at the centre of a circle is 360°, taking the data already used, 360° represents 120 fifth-formers.

$\therefore$ 1 fifth-former is represented by 3°
$\therefore$ 60 fifth-formers are represented by 180°
36 fifth-formers are represented by 108°
18 fifth-formers are represented by 54°
6 fifth-formers are represented by 18°

Measures of averages of individual and discrete data

At this level you are concerned only with *discrete* data. When the data being investigated can take only certain exact values then it is said to be discrete, e.g. the number of goals scored in a football match, the number of stitches in a knitting pattern, the number of children in a class, etc. The values of a discrete variable are normally whole numbers, but this is not always the case. For instance, if we are considering shoe sizes, the variable could take the values 3, $3\frac{1}{2}$, 4, $4\frac{1}{2}$, etc.

An important measure of a set of data is the average, which is a measure of the centre of the data when it has been arranged into a frequency distribution.

There are quite a few different types of average. At this stage we are concerned with three – the mean, the median and the mode.

Mean

The arithmetic mean (or simply, the mean) is what is commonly called the *average*. It is evaluated by summing all the variates and dividing by the total number of variates.

Example: Find the mean of 23, 27, 25, 26, 28, 21.

There are 6 variates, so mean $= \dfrac{23 + 27 + 25 + 26 + 29 + 21}{6}$

$= \dfrac{151}{6} = 25\dfrac{1}{6}$

Using the data of the dice at the start of this module:

$$\text{mean} = \frac{(5 \times 1) + (4 \times 2) + (2 \times 3) + (3 \times 4) + (3 \times 5) + (3 \times 6)}{5 + 4 + 2 + 3 + 3 + 3}$$

$$= \frac{5 + 8 + 6 + 12 + 15 + 18}{20} = \frac{64}{20} = 3.2$$

NOTE: Each variate is multiplied by its own frequency and all the products are added together. The sum is divided by the sum of the frequencies to obtain the mean.

Median

The median of a set of data is the middle value when the data has been arranged in either ascending or descending order. If the data contains an odd number of variates, the median is the *middle* value. If the data contains an even number of variates, the median is the *mean* of the two middle values.

Example 1: Find the median of 7, 2, 6, 8, 2, 5, 3.
The data is arranged in ascending order (just as good to use descending order) *i.e.* 2, 2, 3, 5, 6, 7, 8 (or 8, 7, 6, 5, 3, 2, 2).
Clearly the middle value is 5, therefore the median is 5.

Example 2: Find the median of 7, 2, 6, 8, 2, 5, 3, 4.
Arrange in order *i.e.* 2, 2, 3, 4, 5, 6, 7, 8. Middle two values are 4 and 5,

$\therefore$ median $= \dfrac{4 + 5}{2} = 4.5$

Mode

The mode of a set of data is that variate which occurs the most.

Example 1: 2, 2, 5, 5, 5, 6, 7, 7, 8
Since 5 occurs the most, 5 is the mode.

Example 2: 1, 1, 1, 6, 9, 9, 9, 11, 12
This has 2 modes, 1 and 9, since both occur 3 times.

The arithmetic mean is the most common measure used. It is easy to understand and easy to calculate. Its main disadvantage is that every variate has to be taken into account, even extreme values.

The median is easy to understand and is not affected by extreme values. It is also useful in dealing with qualitative data. The main disadvantage is that it is not suitable for further arithmetic calculations.

The mode also is easy to understand and is not affected by extreme values. Its disadvantage as with the median is that it is not suitable for further arithmetic calculations and it is not widely used.

Probability

This is a measure of the likelihood of some event happening. If an event can succeed in a ways and fail in b ways, then

$$P(\text{success}) = \frac{a}{a+b} \qquad P(\text{failure}) = \frac{b}{a+b}$$

i.e. $P(\text{success}) = \dfrac{\text{Number of times an event can be successful}}{\text{Total number of times event can occur}}$

$P(\text{failure}) = \dfrac{\text{Number of times an event can fail}}{\text{Total number of times event can occur}}$

Probabilities can be found by recording the findings of experiments, or by calculations as in games of chance.

In any problem involving probability, the answer will always be some value from 0 to 1 inclusive.

P (certainty) = 1 P (impossibility) = 0

If today is Tuesday, P (tomorrow is Wednesday) = 1
and P (tomorrow is Friday) = 0

If we toss a fair dice, P (obtaining a 4) = $\frac{1}{6}$,

i.e. one successful outcome in a total of 6 possible events.

If we toss a drawing-pin, P (landing on its back) can be found only by doing an experiment a large number of times and recording how many successes occur. Care must be taken in some problems when probabilities are not equally likely. The sum of the probabilities of all the outcomes will always be 1.

If the probability of a girl passing an exam is 0.6, the probability of failing is $1 - 0.6 = 0.4$.

Example: If a packet of gums contains 4 black, 3 red, 2 orange, 5 yellow and 6 green gums, what is the probability of selecting **a** a yellow gum, **b** not a yellow gum?

There are $4 + 3 + 2 + 5 + 6 = 20$ gums.

a P (yellow) $\frac{5}{20} = \frac{1}{4}$ **b** P (not yellow) $= 1 - \frac{1}{4} = \frac{3}{4}$

NOTE: P (not yellow) could be obtained by finding sum of non yellow, *i.e.* $4 + 3 + 2 + 6 = 15$ and finding by P (not yellow) $= \frac{15}{20} = \frac{3}{4}$

Intermediate Level

Frequency distributions

Frequency distributions can be formed for ungrouped data or grouped data, by using tally marks as shown in the following examples.

Example 1: Suppose the goals scored in the First Division of the Football League were as follows: 0, 1, 2, 1, 3, 2, 0, 0, 4, 2, 1, 1, 2, 1, 4, 1, 0, 0, 1, 0.

Make a table of three columns. The first is the variable under consideration (in this case, goals), the second is the tally, the third is the frequency. So the first column of goals goes from 0 to 4. A vertical stroke is made in the tally column for each of the variates given above. After every fourth vertical stroke, a horizontal one is made to record blocks of five. Count the tallies to get the frequency of each variate.

Goals	*Tally*	*Frequency*
0	~~1111~~ 1	6
1	~~1111~~ 11	7
2	1111	4
3	1	1
4	11	2

The data is ungrouped and each variate is denoted by its own frequency.

In some examples, we condense the information by grouping certain variates together in order to obtain a frequency distribution that is more easily interpreted. In such examples, the data is grouped into class intervals. The number of class intervals considered would normally be about 10. In some examples the class intervals would be given, in others they would need to be determined by you.

Example 2: Suppose 60 people score the following marks in a Maths examination, where marks can be from 0 to 100.

81	36	40	45	92	18	27	46	37	62
12	45	91	26	27	36	45	52	53	31
8	49	61	62	56	11	71	77	63	68
71	48	64	71	58	57	58	59	62	49
36	14	52	63	42	58	60	57	39	48
80	48	58	66	68	54	59	48	46	53

A convenient class interval could well be in 10s, say 0-9, 10-19, 20-29, etc.

Class interval	*Tally*	*Frequency*
0-9	1	1
10-19	1111	4
20-29	111	3
30-39	~~1111~~ 1	6
40-49	~~1111~~ ~~1111~~ 111	13
50-59	~~1111~~ ~~1111~~ 1111	14
60-69	~~1111~~ ~~1111~~ 1	11
70-79	1111	4
80-89	11	2
90-99	11	2

The information is much more manageable in this form, although of course exact detail has now been lost.

Histograms

This is a method of pictorial representation and is similar to a vertical bar graph without spaces between the bars, but the fundamental difference is that it is the area of each bar that is proportional to the frequency of a variate rather than the length of a bar. At this level only histograms of equal widths will be considered and therefore the length of each bar will be proportional to the frequency of a variate.

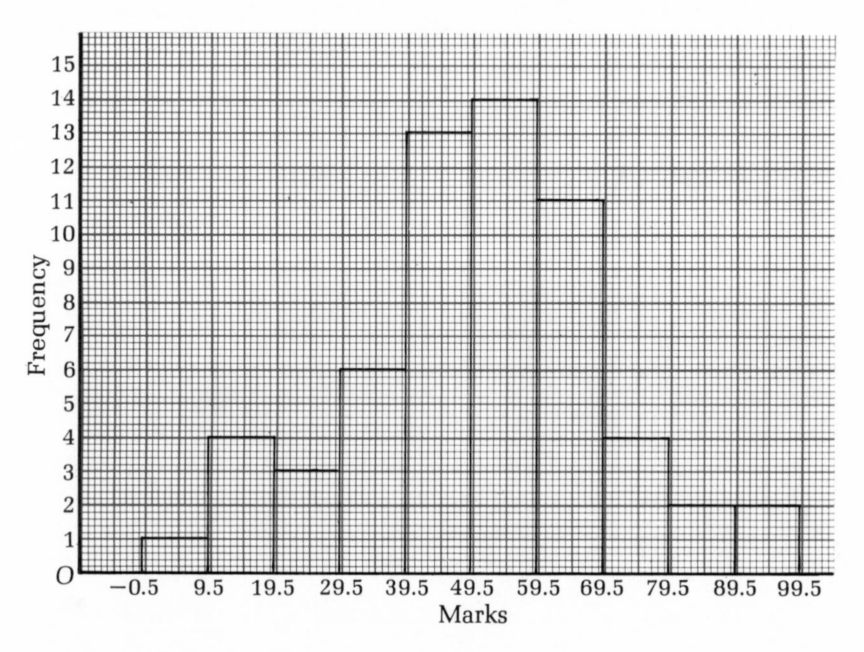

NOTE: With the examination data above, the class intervals were 10 marks so that the class boundaries are in this case −0.5, 9.5, 19.5, 29.5, etc.

For discrete data, such as in the earlier example on dice throwing, denote the values of the variates at the centre of each bar on the horizontal axis when constructing a histogram.

Example:

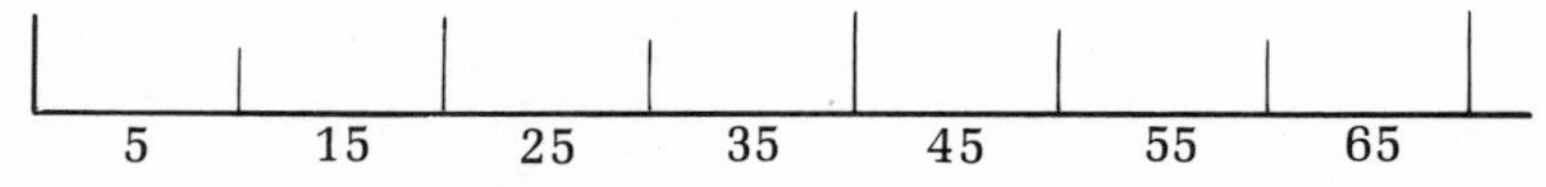

Simple combined probabilities

Problems on combined probability can often be solved more easily by diagrammatic methods. The first method is called a possibility diagram or a sample space.

Example 1: When a coin is tossed the possibility diagram is simply:

Head Tail.

If two coins are tossed, then the possibility diagram is as shown:

		First coin	
		H	T
Second coin	H	HH	TH
	T	HT	TT

4 possible outcomes

$P(2H) = \frac{1}{4}$ $P(2T) = \frac{1}{4}$ $P(HT \text{ or } TH) = \frac{2}{4} = \frac{1}{2}$

Example 2: If a spinner with five equal sectors, marked 1, 2, 3, 4 and 5, is spun then the possibility diagram is:

1, 2, 3, 4, 5

If a second identical spinner is spun and the scores added together, the possibility diagram is:

		1st spinner				
		1	2	3	4	5
	1	2	3	4	5	6
	2	3	4	5	6	7
2nd spinner	3	4	5	6	7	8
	4	5	6	7	8	9
	5	6	7	8	9	10

then $P \text{ (score of 7)} = \frac{4}{25}$

$P \text{ (score} > 8) = \frac{3}{25}$

$P \text{ (score} \leqslant 4) = \frac{6}{25}$

$P \text{ (score of 1)} = 0$

Probability tree diagrams

Many problems can be made easier by drawing a probability tree diagram for combining probabilities.

Example: A box contains 6 red marbles and 4 white marbles. A marble is taken out at random and not replaced. A second marble is taken out. Find the probabilities that the two marbles are **a** both red, **b** both white, **c** different colours.

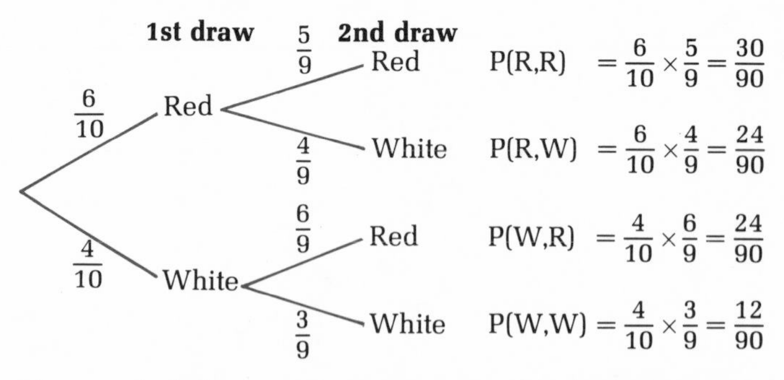

The probabilities are written on the branches of the tree, and at any stage the probability for any situation is obtained by multiplying these single probabilities.

a P(both red) $= \frac{6}{10} \times \frac{5}{9} = \frac{30}{90} = \frac{1}{3}$

b P(both white) $= \frac{4}{10} \times \frac{3}{9} = \frac{12}{90} = \frac{2}{15}$

c P(different colours) $= 1 - \frac{1}{3} - \frac{2}{15} = 1 - \frac{5+2}{15} = 1 - \frac{7}{15} = \frac{8}{15}$

or $\frac{6}{10} \times \frac{4}{9} + \frac{4}{10} \times \frac{6}{9} = \frac{24}{90} + \frac{24}{90} = \frac{48}{90} = \frac{8}{15}$

Clearly the probability for a third withdrawal is obtained by extending the diagram.

Be careful in reading the question. If the marble were replaced, it would be as follows.

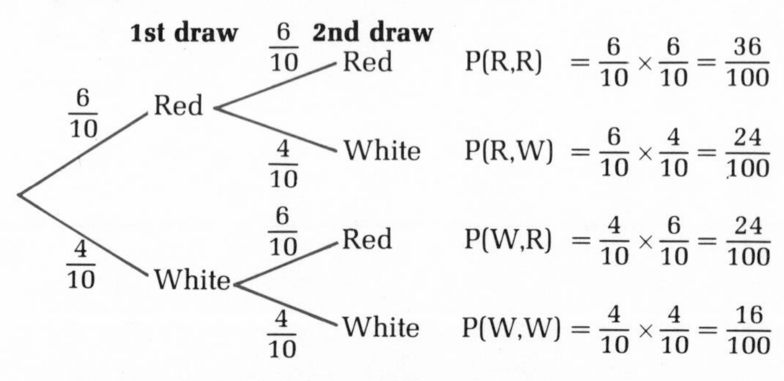

NOTE: At any stage, total probability is always 1. In the last case $\frac{36}{100} + \frac{24}{100} + \frac{24}{100} + \frac{16}{100} = \frac{100}{100} = 1.$

Higher Level

Continuous variables

Some variables, such as length or mass, cannot be measured exactly as a discrete variable and are said to be continuous variables. The averages – mean, median and mode – can be found for continuous and grouped data as follows.

Using the grouped data for the examination marks earlier in the module:

The *modal group* is 50-59 since it occurs 14 times.

The *median* can be estimated from grouped data
a by calculation.
b from a cumulative frequency curve (see page 56).

The median is the middle value of a set of data and for n variates is the $\left(\frac{n+1}{2}\right)$th value.

In the example n is 60. Therefore we need $\left(\frac{60+1}{2}\right)$th value, *i.e.* $30\frac{1}{2}$th.

Summing the frequencies, the $30\frac{1}{2}$th value occurs in the 50-59 class interval.

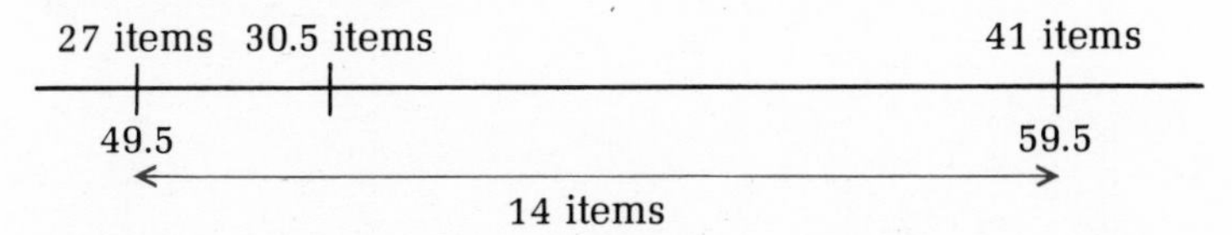

There are 27 items up to 49.5 and 41 items up to 59.5.

$$\text{Estimate} = 49.5 + \frac{(30.5 - 27)10}{14} = 49.5 + 2.5 = 52$$

The *mean* is estimated by taking the midpoint of each class interval to be representative of that interval.

NOTE: In this case, the mean will not be exact but a good estimate.

Midpoint	*Frequency*	*Midpoint × Frequency*
4.5	1	4.5
14.5	4	58
24.5	3	73.5
34.5	6	207
44.5	13	578.5
54.5	14	763
64.5	11	709.5
74.5	4	298
84.5	2	169
94.5	2	189
	60	3050

$$\text{Estimated mean} = \frac{3050}{60} = 50.833 = 50.8 \text{ (3 s.f.)}$$

The same techniques are applied for continuous data.

The range

This is a measure of the spread of a set of variates. It is the difference between the highest and lowest variates.

In the example of the examination marks, the lowest mark is 8, and the highest mark is 92.

Therefore range = 92 − 8 = 84.

Cumulative frequency diagrams

The cumulative frequency is the total frequency up to a particular variate or class boundary.

Again we can use the example on examination marks.

The cumulative frequencies are plotted against the upper class boundaries. The points are joined by a smooth curve to give an *ogive* or cumulative frequency curve.

NOTE: The starting point would be at −0.5.

Upper Class Boundary	*Cumulative Frequency*
9.5	1
19.5	5
29.5	8
39.5	14
49.5	27
59.5	41
69.5	52
79.5	56
89.5	58
99.5	60

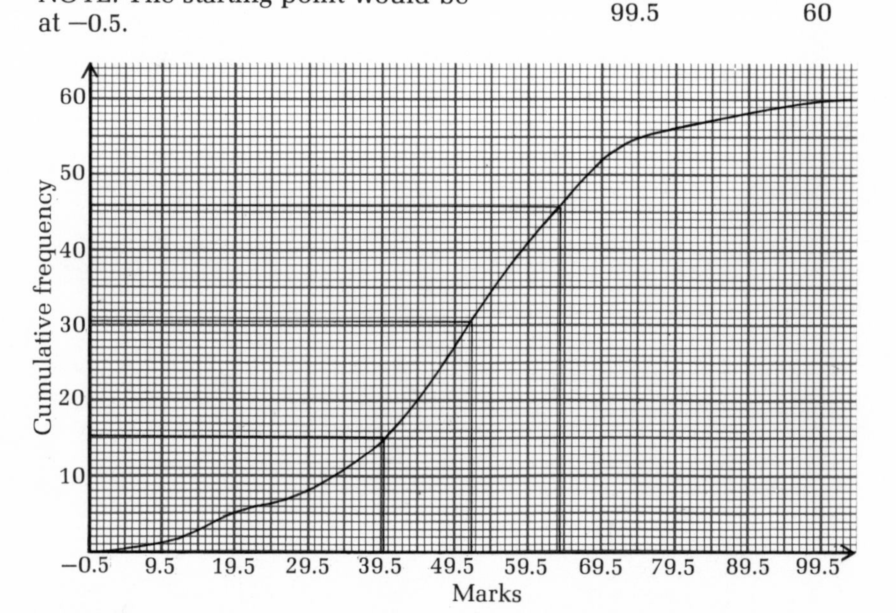

Quartiles

The three values which split a distribution into four equal portions are called quartiles.

For n values, the lower quartile Q_1 is $\frac{1}{4}(n + 1)$th value

the median quartile Q_2 is $\frac{1}{2}(n + 1)$th value

the upper quartile Q_3 is $\frac{3}{4}(n + 1)$th value

On the diagram, $Q_1 = 40 \quad Q_2 = 52 \quad Q_3 = 64$

The interquartile range is a measure of the spread of results and this is not affected by extreme results since it is purely dependent on the middle half of the frequency distribution. It is defined by $Q_3 - Q_1$, i.e. $64 - 40$ in last example, which is 24.

1 Algebra

Foundation Level

The use of letters for generalised numbers, basic operations, substitution of numbers, multiplication and division of algebraic quantities.

Intermediate Level

Transformation (or transposition) of simple formulae, directed numbers, use of brackets, extraction of common factors, use of integral indices, construction and solution of simple equations.

Higher Level

Harder transformation of formulae, positive and negative fractional indices, products of two binomial expressions, factorisation, manipulation of algebraic fractions (addition and subtraction, multiplication and division), solution of fractional equations, of simultaneous linear equations in two unknowns, of quadratic equations.

Foundation Level

The use of letters for generalised numbers

Algebra uses letters and symbols as well as the numbers that are used in arithmetic. In arithmetic we are dealing with known numbers, but in algebra we generalise situations. For instance Rahat might get £3 per week for pocket money, Peter might be 16 years old, Jane might take size 12 in dresses, Rajput might be 1.83m tall. It could well be the case that we do not have this particular set of information, but other information that would enable us to find it. Therefore we could generalise these statements as follows, by using letters instead of actual numbers.

i.e. Rahat might get £x per week.
Peter might be y years old.
Jane might take size a in dresses
Rajput might be b m tall.

We can use what letters or symbols we wish – the choice is arbitrary.

Basic operations

In addition or subtraction, only 'like terms' can be collected together.

Example 1: $7x + 3x = 10x$

Example 2: $5y - 3y = 2y$

Example 3: $2x + 6y + 9x - 4y = 11x + 2y$ $(2x + 9x = 11x, 6y - 4y = 2y)$

Example 4: $3a + 2b - 6c + 7a - 3b - 9c = 10a - b - 15c$

Example 5: $x + 3y + 7t$ has three unlike terms and cannot be further simplified.

Example 6: $x^2 + 2x^2 + 5x^2 - 3x^2 = 5x^2$

Substitution of numbers in algebraic statements

A letter might well have a particular value in a statement. If so, we substitute for that letter and evaluate the value of the statement.

Example: If $x = 2$, $y = 3$, find the values of

a $3x - 2y$ **b** $5x^2 + y$ **c** $2x + y^2$ **d** $\frac{5x}{3y}$.

a $3(2) - 2(3) = 6 - 6 = 0$

b $5(2)^2 + 3 = 5 \times 4 + 3 = 20 + 3 = 23$

c $2(2) + (3)^2 = 4 + 9 = 13$

d $\frac{5(2)}{3(3)} = \frac{10}{9} = 1\frac{1}{9}$

Multiplication and division of algebraic quantities

$a \times b$ is written as ab.

$\therefore$ xyz means $x \times y \times z$

$5x \times 3y$ means $5 \times x \times 3 \times y = 5 \times 3 \times x \times y = 15xy$

$\therefore 20a \times 3b \times 2c = 120abc$

$a \div b$ is written as $\frac{a}{b}$

$\therefore 8a \div 2b = \frac{8a}{2b} = \frac{4a}{b}$ (cancel by factor 2 in numerator and denominator)

$6x^2 \times 2x^5 = 6 \times x \times x \times 2 \times x \times x \times x \times x \times x = 12x^7$

$30y^3 \div 6y = \frac{30y^3}{6y} = 5y^2$ (remember Indices Rules in Module 1)

Example: If $x = 1$, $y = 2$, $z = 0$, evaluate

a xyz **b** $\frac{xy + z}{y^2}$ **c** $5x \times 3y$.

a $xyz = 1 \times 2 \times 0 = 0$

b $\frac{xy + z}{y^2} = \frac{1 \times 2 + 0}{2^2} = \frac{2}{4} = \frac{1}{2}$

c $5x \times 3y = 5 \times 1 \times 3 \times 2 = 30$

Intermediate Level

Transformation or transposition of simple formulae

If you have some formula such as $y = ax$, and you wish to express x on its own, divide both sides of the formula to get $\frac{y}{a} = \frac{ax}{a}$,

i.e. $\frac{y}{a} = x$. This is called *transforming* the formula, or *transposing* the formula, or making x the subject of the formula.

Example 1: If $y = u + at$, make t the subject of the formula.

Firstly get rid of terms not containing t from the RHS (right-hand side).

$y - u = at$ (taking u from each side)

$\frac{y - u}{a} = t$ (dividing each side by a)

Example 2: If $I = \frac{PRT}{100}$, make R the subject of the formula.

$100I = PRT$ (multiplying both sides by 100)

$\frac{100I}{PT} = R$ (dividing both sides by PT)

Example 3: If $C = 2\pi r$, make r the subject of the formula.

$\frac{C}{2\pi} = r$ (dividing both sides by 2π)

Example 4: If $s = \frac{1}{2}at^2$, make a the subject of the formula.

$2s = at^2$ (multiplying both sides by 2)

$\frac{2s}{t^2} = a$ (dividing both sides by t^2).

Directed numbers

The rules are those for arithmetic (see Module 3).

Examples: $-3x \times -2y = 6xy$ $\quad 3x \times 2y = 6xy$

$3x \times -2y = -6xy$ $\quad -3x \times 2y = -6xy$

Use of brackets

Each term inside the bracket must be multiplied by the number or letter outside the bracket.

Example 1: $2(x + 4) = 2 \times x + 2 \times 4 = 2x + 8$

Example 2: $3(2x - 4y) = 3 \times 2x + 3 \times - 4y = 6x - 12y$

Example 3: $-6(3x + 2y) = -6 \times 3x + -6 \times 2y = -18x - 12y$

Example 4: $-4(2x - y) = -4 \times 2x + -4 \times -y = -8x + 4y$

Extraction of common factors

This is basically the reverse of the previous example for bracket removal.

$2x + 8 = 2(x + 4)$ $\quad$ 2 is a common factor of $2x$ and 8.

$6x - 12y = 6(x - 2y)$ $\quad$ 6 is a common factor of $6x$ and $-12y$.

$-3p + 6q = -3(p - 2q)$ or $3(2q - p)$

$a^2 + 6ab = a(a + 6b)$

Check that if you remove the brackets on the RHS you get what you started with.

Use of integral indices

Example 1: $x^3 \times x^5 = x^{3+5} = x^8$

Example 2: $x^{-3} \times x^5 = x^{-3+5} = x^2$

Example 3: $x^7 \div x^9 = x^{7-9} = x^{-2}$

Example 4: $(x^2)^3 = x^2 \times x^2 \times x^2 = x^6$

Example 5: $(x^2y)^4 = x^8y^4$

Remember, $a^{-1} = \frac{1}{a}$ and $a^{-2} = \frac{1}{a^2}$.

In general: $a^{-P} = \frac{1}{a^P}$.

Construction and solution of simple equations

To translate information into algebriac expressions, you generally introduce some letter for the quantity that has to be found. Check when the equation has been found that the units of each side of the equation are the same.

Example 1: I think of a number, double it and add 4. The result is 12. What is the number?

Let x be the number, double it, *i.e.* 2x.

$\therefore 2x + 4 = 12$

$2x + 4 - 4 = 12 - 4$ (taking 4 from both sides)

$2x = 8$

$x = \frac{8}{2} = 4$ (dividing both sides by 2)

$\therefore$ the number is 4.

Example 2: A man buys 10 newspapers. Some papers cost 20p each and the rest 22p each. He pays £2.08 for the papers. How many does he buy @ 20p each.

Let number of papers at 20p each be x.
Let number of papers at 22p each be (10 − x).
Cost of papers @ 20p = 20x p
Cost of papers @ 22p = 22(10 − x) p
Cost of papers = £2.08 = 208p (units must be consistent)

$\therefore 20x + 22(10 - x) = 208$

$20x + 220 - 22x = 208$

$220 - 208 = 22x - 20x$ (collecting like terms)

$12 = 2x$

$x = \frac{12}{2} = 6$

$\therefore$ he buys 6 papers at 20p each.

Example 3: The three sides of a triangle are such that two sides are respectively 5 cm and 8 cm greater than the third side. Find the length of the third side if the perimeter of the triangle is 0.73 m.

Let x cm be the length of the third side.

$\therefore$ other two sides are (x + 5)cm and (x + 8)cm

Perimeter = 0.73 m = 73 cm (make sure the units are consistent)

Then $x + x + 5 + x + 8 = 73$

$3x + 13 = 73$

$3x = 73 - 13$ (taking 13 from each side)

$3x = 60$

$x = \frac{60}{3} = 20$

$\therefore$ third side is 20 cm

Higher Level

Harder examples on transformation of formulae

There are no set rules and each question must be done on its own merit.

Example 1: If $t = 2\pi\sqrt{\frac{l}{g}}$, make l the subject of the formula.

$t^2 = \frac{4\pi^2 l}{g}$ (squaring both sides)

$gt^2 = 4\pi^2 l$ (multiplying both sides by g)

$\frac{gt^2}{4\pi^2} = l$ (dividing both sides by $4\pi^2$)

Example 2: If $A = \pi r\,(r + 2h)$, make h the subject of the formula.

$\frac{A}{\pi r} = r + 2h$ (dividing both sides by πr)

$\frac{A}{\pi r} - r = 2h$ (get $2h$ on its own)

$\therefore \frac{1}{2}\left(\frac{A}{\pi r} - r\right) = h$ (dividing both sides by 2)

Example 3: If $a = \frac{5b + 2}{b + 3}$, make b the subject of the formula.

$a(b + 3) = 5b + 2$ (get rid of fractions, multiplying both sides by $b + 3$)

$ab + 3a = 5b + 2$ (simplifying brackets)

$ab - 5b = 2 - 3a$ (all terms with b to one side, rest to other side)

$b(a - 5) = 2 - 3a$ (factorising, b is common)

$b = \frac{2 - 3a}{a - 5}$ (dividing both sides by $a - 5$)

Positive and negative fractional indices

The rules for indices still apply in the operation of fractional indices.

What does $a^{\frac{1}{2}}$ mean?

In arithmetic we know $2 \times 2 = 4$ — 2 is the sq. root of 4

$7 \times 7 = 49$ — 7 is the sq. root of 49

$10 \times 10 = 100$ — 10 is the sq. root of 100.

Since $a^{\frac{1}{2}} \times a^{\frac{1}{2}} = a^{\frac{1}{2}+\frac{1}{2}} = a^1$, then $a^{\frac{1}{2}}$ is the square root of a.

Likewise $a^{\frac{1}{3}}$ is the cube root of a, i.e. $\sqrt[3]{a}$.

In general: $a^{\frac{p}{q}} = \sqrt[q]{a^p}$.

Hence $a^{-\frac{1}{2}}$ is $\frac{1}{a^{\frac{1}{2}}} = \frac{1}{\sqrt{a}}$.

Products of two binomial expressions

A binomial expression consists of two terms, e.g. $2x + 1$, $p - 3q$, $3a + 5b$, or $2m - 7a$.

The product of two binomial expressions such as $(x + y) \times (a + b)$ is evaluated by multiplying each term of the second bracket by each term of the first bracket,

i.e. $(x + y)(a + b) = x(a + b) + y(a + b) = ax + bx + ay + by$

We can omit the middle step by considering the individual products shown by the answers below.

$$(x + y)(a + b) = ax + bx + ay + by.$$

Example 1: $(2x + 1)(3x - 2) = 6x^2 - 4x + 3x - 2$
$= 6x^2 - x - 2$

Example 2: $(3y - 4)(4y - 5) = 12y^2 - 15y - 16y + 20$
$= 12y^2 - 31y + 20$

Be careful with the signs of the products.
Remember, the product of 2 like signs is always positive.
Similarly, the product of 2 unlike signs is always negative.

Factorisation

This process is the reverse process of the previous one. An algebraic expression is expressed as the product of factors.

We need to consider two separate cases:

Factorising by grouping

If we have $ax + bx + ay + by$, the expression is firstly considered by taking these terms in pairs:

$$(ax + bx) + (ay + by) = x(a + b) + y(a + b).$$

i.e. x is a common factor of the first pair of terms, and y is a common factor of the second pair of terms.

Now, note that $(a + b)$ is a common factor of $x(a + b) + y(a + b)$; hence $x(a + b) + y(a + b) = (a + b)(x + y)$.

Check by expanding the binomial product.

Further examples are shown.

Example 1: $6xy - 9y - 8x + 12 = 3y(2x - 3) - 4(2x - 3)$
$= (2x - 3)(3y - 4)$

Example 2: $2xy - 4xz + 3y - 6z = 2x(y - 2z) + 3(y - 2z).$
$= (y - 2z)(2x + 3)$

Sometimes we need to rearrange the given expression.

Example: $pq + rs + pr + qs$

i.e. $pq + pr + rs + qs = p(q + r) + s(r + q) = (q + r)(p + s)$

NOTE: $(q + r)$ is the same as $(r + q)$

Factors of quadratic expressions

A quadratic expression is of the form $ax^2 + bx + c$ where a, b and c are integers but a cannot $= 0$. For instance, $3x^2 + 2x - 7$, $x^2 + 5x$, $2x^2 - 7$ are quadratic expressions, *i.e.* there must be an expression where the highest power used is 2.

Many quadratic expressions can be factorised to give the product of two binomial terms.

There are three possible types.

Type A

If we have type $ax^2 + bx$, then $x(ax + b)$ is the factorised form, x is a common factor.

Example 1: $3x^2 - 2x = x(3x - 2)$ *Example 2:* $4x^2 + 16x = 4x(x + 4)$

Type B

If we have type $a^2x^2 - b^2$, then $(ax - b)(ax + b)$ is the factorised form. This is a special case called the difference of two squares $X^2 - Y^2 = (X - Y)(X + Y)$.

Check the expansion of the RHS gives the LHS.

Example 1: $x^2 - 4 = x^2 - 2^2 = (x - 2)(x + 2)$

Example 2: $9y^2 - 16x^2 = (3y)^2 - (4x)^2 = (3y - 4x)(3y + 4x)$

Example 3: $8a^2 - 200 = 8(a^2 - 25) = 8(a^2 - 5^2) = 8(a - 5)(a + 5)$

Type C

The more general type of $ax^2 + bx + c$. These are best factorised by trial and error.

Earlier we showed that $(2x + 1)(3x - 2) = 6x^2 - x - 2$. You need to factorise $6x^2 - x - 2$ to obtain $(2x + 1)(3x - 2)$. Note that 2x and 3x are factors of $6x^2$, and 1 and -2 are factors of -2. The method by trial and error needs us to find all possible factors of $6x^2$ and -2.

Factors of $6x^2$	Factors of -2
6x and x	1 and -2
3x and 2x	-1 and 2

Test all possible combinations $(6x + 1)(x - 2)$:

$(6x + 1)(x - 2) = 6x^2 - 12x + x - 2 = 6x^2 - 11x - 2$ wrong
$(6x - 2)(x + 1) = 6x^2 + 6x - 2x - 2 = 6x^2 + 4x - 2$ wrong
$(6x - 1)(x + 2) = 6x^2 + 12x - x - 2 = 6x^2 + 11x - 2$ wrong
$(6x + 2)(x - 1) = 6x^2 - 6x + 2x - 2 = 6x^2 - 4x - 2$ wrong
$(3x + 2)(2x - 1) = 6x^2 - 3x + 4x - 2 = 6x^2 + x - 2$ wrong
$(3x + 1)(2x - 2) = 6x^2 - 6x + 2x - 2 = 6x^2 - 4x - 2$ wrong
$(3x - 1)(2x + 2) = 6x^2 + 6x - 2x - 2 = 6x^2 + 4x - 2$ wrong
$(3x - 2)(2x + 1) = 6x^2 + 3x - 4x - 2 = 6x^2 - x - 2$ correct.

By trying all possible solutions, you eventually find the correct one. With practice you will be able to find the correct solution more quickly.

Check that you could factorise the following.

Example 1: $x^2 + 7x + 10 = (x + 5)(x + 2)$

Example 2: $x^2 - x - 12 = (x - 4)(x + 3)$

Example 3: $6y^2 + y - 35 = (3y - 7)(2y + 5)$

Example 4: $2a^2 - 5a - 3 = (2a + 1)(a - 3)$

Manipulation of algebraic fractions

The rules for ordinary arithmetical fractions apply to algebraic fractions also.

Addition and subtraction

Example 1: Add $\frac{x}{5}$ to $\frac{x-3}{4}$.

$$\frac{x}{5} + \frac{x-3}{4} = \frac{4x + 5(x-3)}{20} \quad \text{(find lowest common denominator and obtain equivalent fraction)}$$

$$= \frac{4x + 5x - 15}{20} \quad \text{(simplify numerator)}$$

$$= \frac{9x - 15}{20}$$

$$= \frac{3(3x - 5)}{20} \quad \text{(factorise numerator)}$$

No common factor to numerator and denominator, so cannot be simplified further.

Example 2: $$\frac{2x+1}{6} - \frac{x-2}{2} = \frac{(2x+1) - 3(x-2)}{6}$$

$$= \frac{2x + 1 - 3x + 6}{6} = \frac{7 - x}{6}$$

NOTE: Be careful with the signs here.

Example 3: $$\frac{4}{x-2} + \frac{3}{x+1} = \frac{4(x+1) + 3(x-2)}{(x-2)(x+1)}$$

$$= \frac{4x + 4 + 3x - 6}{(x-2)(x+1)} = \frac{7x - 2}{(x-2)(x+1)}$$

Multiplication and division

Example 1: $$\frac{3x}{4} \times \frac{8y}{x^2} = \frac{3 \times x \times 8 \times y}{4 \times x \times x} = \frac{24xy}{4x^2} = \frac{6y}{x}$$

or $$\frac{3\cancel{x}}{_1\cancel{4}} \times \frac{^2\cancel{8}y}{x^{\cancel{2}}} = \frac{6y}{x}$$ (by cancelling the factors of 4 and x)

Example 2: $$\frac{6xy}{5a} \div \frac{3y}{10b} = \frac{^2\cancel{6}x\cancel{y}}{_1\cancel{5}a} \times \frac{^2\cancel{10}b}{_1\cancel{3y}} = \frac{4bx}{a}$$

Example 3: $$\frac{2x}{x-2} \div \frac{3x^2}{5x-10} = \frac{2x}{x-2} \times \frac{5x-10}{3x^2}$$

$$= \frac{2\cancel{x}}{\cancel{x-2}} \times \frac{5\cancel{(x-2)}}{3x^{\cancel{2}}} = \frac{10}{3x}$$

Solution of fractional equations

The way to solve equations containing fractions is by multiplying *each* term of the equation by the lowest common multiple (LCM) of the denominators.

Example 1: Solve $\frac{2x}{5} - \frac{x}{3} = 4$.

Multiply each term of the equation by LCM of 5 and 3, which is 15:

$$15 \times \frac{2x}{5} - 15 \times \frac{x}{3} = 15 \times 4$$

i.e. $3 \times 2x - 5 \times x = 60$ (cancelling factors)

$$6x - 5x = 60$$
$$x = 60$$

Example 2: Solve $\frac{15}{x} = 5$.

Multiply each term by x (LCM):

$$x \times \frac{15}{x} = x \times 5$$
$$15 = 5x$$

i.e. $\frac{15}{5} = x$

$$3 = x$$

Example 3: $\frac{5}{x-2} = 2$

Multiply each term by $(x - 2)$:

$$(x-2) \times \frac{5}{x-2} = 2(x-2)$$
$$5 = 2x - 4$$
$$5 + 4 = 2x$$
$$9 = 2x$$
$$4.5 = x$$

Example 4: $\frac{x+2}{3} - \frac{x-1}{5} = 2$

Multiply each term by 15:

$15 \times \frac{(x+2)}{3} - 15 \times \frac{(x-1)}{5} = 15 \times 2$ Note the use of brackets.

$$5(x+2) - 3(x-1) = 30$$
$$5x + 10 - 3x + 3 = 30$$
$$2x = 30 - 10 - 3$$
$$2x = 17$$
$$x = 8.5$$

Solution of simultaneous linear equations in two unknowns

A linear equation is of the form $ax + by = c$ where a, b, c are constants. In simultaneous linear equations we shall be concerned with two such equations where x and y are unknown quantities. We need to find the value of x and the value of y which satisfy both equations.

The easiest way to solve such equations is by eliminating one of the unknowns.

Suppose the two equations are $2x + 3y = 16$ and $3x - 2y = 11$ and you want to find the values of x and y that satisfy both.

Firstly denote the equation $2x + 3y = 16$ ①

$3x - 2y = 11$ ②

The method entails making the coefficients of x or y (irrespective of signs) the same in both equations.

Thus, ① × 2 becomes $4x + 6y = 32$ ③
② × 3 becomes $9x - 6y = 33$ ④

The terms in y can be eliminated by adding ③ and ④,

i.e. $13x = 65$

$$\therefore x = \frac{65}{13} = 5$$

To find y, substitute $x = 5$ in either ① or ②.

Suppose we use ①; then $2(5) + 3y = 16$

$$10 + 3y = 16$$
$$3y = 16 - 10$$
$$3y = 6$$
$$y = \frac{6}{3} = 2$$

$\therefore x = 5$, $y = 2$ are the required solutions.

Check in ② LHS $= 3(5) - 2(2) = 15 - 4 = 11 =$ RHS

Example 1: Solve for x and y:

$7x + 2y = 13$ ①
$3x + 4y = 15$ ②
① × 2 $14x + 4y = 26$ ③
③ − ② $11x = 11$

$$x = \frac{11}{11} = 1$$

Substitute in ① $7(1) + 2y = 13$

$$7 + 2y = 13$$
$$2y = 13 - 7$$
$$2y = 6$$
$$y = \frac{6}{2} = 3$$
$$\therefore x = 1,\ y = 3$$

Check in ② LHS $= 3(1) + 4(3) = 3 + 12 = 15 =$ RHS

Example 2: Solve for x and y:

$3x - 2y = 12$ ①
$2x - 5y = 19$ ②
① × 5 $15x - 10y = 60$ ③
② × 2 $4x - 10y = 38$ ④
③ − ④ $11x = 22$ (*N.B.* $-10y - -10y = -10y + 10y = 0$)

$$x = \frac{22}{11} = 2$$

Substitute in ① $3(2) - 2y = 12$

$$6 - 2y = 12$$
$$6 - 12 = 2y$$
$$-6 = 2y, \quad \text{i.e. } y = -\frac{6}{2} = -3$$
$$\therefore\ x = 2,\ y = -3$$

Check in ② LHS $= 2(2) - 5(-3) = 4 + 15 = 19 =$ RHS

Solution of quadratic equations

These are of the type $ax^2 + bx + c = 0$ where a, b, c, are integers except $a \neq 0$.

We use the methods of factorisation, if possible, given earlier in the module.

Type A – if $c = 0$

If $3x^2 - 2x = 0$ then $x(3x - 2) = 0$

If the product of x and $(3x - 2)$ is zero then

either $x = 0$ or $3x - 2 = 0$

$\therefore\ x = 0$ or $3x = 2$

$x = 0$ or $x = \frac{2}{3}$ are the solutions

Type B – if $b = 0$

If $3x^2 - 6 = 0$ then $3x^2 = 6$

$$x^2 = \frac{6}{3} = 2$$
$$x = \pm\sqrt{2}$$

NOTE: When we take the square root of any positive quantity we obtain two answers, one positive and one negative.

BUT: If we had $3x^2 + 6 = 0$ and thus $3x^2 = -6$

It would give $x^2 = \frac{-6}{3} = -2$.

In the system of real numbers, we cannot take the square root of a negative number, so there is no real solution. This type is most unlikely to occur in the examination.

Type C – *if* a, b and c all exist

If $6x^2 - x - 2 = 0$ then $(3x - 2)(2x + 1) = 0$

$\therefore$ either $3x - 2 = 0$ or $2x + 1 = 0$

i.e. $3x = 2$ or $2x = -1$

$$\therefore\ x = \frac{2}{3} \quad \text{or} \quad -\frac{1}{2}$$

The formula for solutions

Not all quadratic expressions can be factorised but there is a formula that can be used to solve quadratic equations.

Consider $ax^2 + bx + c = 0$.

Divide by a: $x^2 + \frac{b}{a}x + \frac{c}{a} = 0 \quad \therefore\ x^2 + \frac{b}{a}x = -\frac{c}{a}$

Complete the square of the LHS. This is done by taking half the coefficient of x, squaring this, and adding to both sides of the equation.

$$x^2 + \frac{b}{a}x + \left(\frac{b}{2a}\right)^2 = \left(\frac{b}{2a}\right)^2 - \frac{c}{a}$$

The LHS is a perfect square.

$$\left(x + \frac{b}{2a}\right)\left(x + \frac{b}{2a}\right) = \frac{b^2}{4a^2} - \frac{c}{a} = \frac{b^2 - 4ac}{4a^2}$$

Take square root of each side.

$$x + \frac{b}{2a} = \frac{\pm\sqrt{b^2 - 4ac}}{2a} \qquad \therefore x = \frac{-b \pm \sqrt{b^2 - 4ac}}{2a}$$

You will not be required to learn the proof but you will need to apply the formula.

Example: Solve $2x^2 + 5x - 8 = 0$ giving the answers to 2 d.p.
Compare to $ax^2 + bx + c = 0$. $a = 2$, $b = 5$, $c = -8$.

$$\therefore x = \frac{-5 \pm \sqrt{5^2 - 4(2)(-8)}}{2(2)}$$

$$= \frac{-5 \pm \sqrt{25 + 64}}{4}$$

$$= \frac{-5 \pm \sqrt{89}}{4}$$

$$= \frac{-5 + 9.4339811}{4} \quad \text{or} \quad \frac{-5 - 9.4339811}{4}$$

$= 1.11$ or -3.61 (to 2 d.p.)

In the examination, the equation may require some prior manipulation before you can apply the formula or indeed before factorising.

Example: Solve $(x + 3)(x - 4) = 5$. Give answers to 3 s.f.
Multiply out LHS.
then

$$x^2 - 4x + 3x - 12 = 5$$
$$x^2 - x - 12 - 5 = 0$$
$$x^2 - x - 17 = 0$$

$\therefore$ compare to $ax^2 + bx + c = 0$ $a = 1$, $b = -1$, $c = -17$

$$\therefore x = \frac{- -1 \pm \sqrt{(-1)^2 - 4(1)(-17)}}{2(1)}$$

$$= \frac{1 \pm \sqrt{1 + 68}}{2}$$

$$= \frac{1 \pm \sqrt{69}}{2}$$

$$= \frac{1 + 8.3066239}{2} \quad \text{or} \quad \frac{1 - 8.3066239}{2}$$

$= 4.65$ or -3.65 (to 3 s.f.)

2 Graphs

Foundation Level
Axes, Cartesian coordinates, interpretation and use of graphs.

Intermediate Level
Plotting graphs of algebriac functions, idea of gradient.

Higher Level
Graphs of harder algebraic functions, inequalities.

Foundation Level

Axes

Information is often represented pictorially by means of a graph. The most general type of graph is a diagram in which two lines, intersecting at right angles, are taken to be the axes of reference. It is normal to draw these axes vertically and horizontally. In many practical instances like in travel graphs or conversion graphs, we need only represent positive values such as distance and time or pounds and francs. In such cases, we would draw the axes as follows.

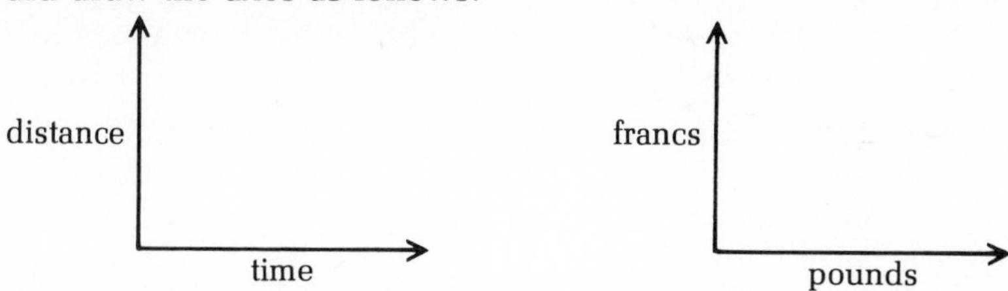

The number of units represented by a unit of length along an axis is called the scale. At this level, the scales to be used will be given in the question.

In some algebraic graphs, we deal with both positive and negative quantities. The axes are extended in both directions as follows.

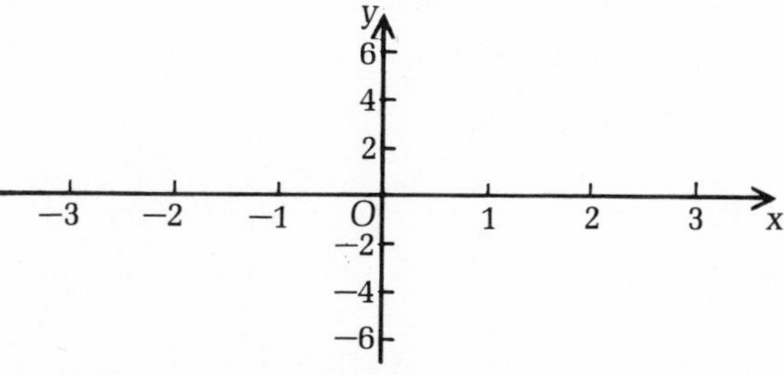

The axes intersect at the origin, *O*. The horizontal axis takes positive values of *x* to the right of the origin and negative values to the left of the origin. The scale for positive and negative values of *x must* be the same.

The vertical axis takes positive values of *y* above the origin and negative values below the origin; again, the scale for these values must be the same. You can of course have different scales for the *x*-values and the *y*-values.

Cartesian coordinates

These are named after the French philosopher and mathematician, René Descartes. While lying on his sick-bed, he noticed a fly walking on the ceiling where there were two cracks. He was able to fix the fly's position by measuring the distances from the two cracks which were approximately at right angles to each other. These distances were represented by ordered pairs, called cartesian coordinates, given in the form (x, y). The x-value, i.e. the distance along the horizontal axis, is *always* given first; the y-value, the distance along the vertical axis, is given second.

The following graph shows how this works.

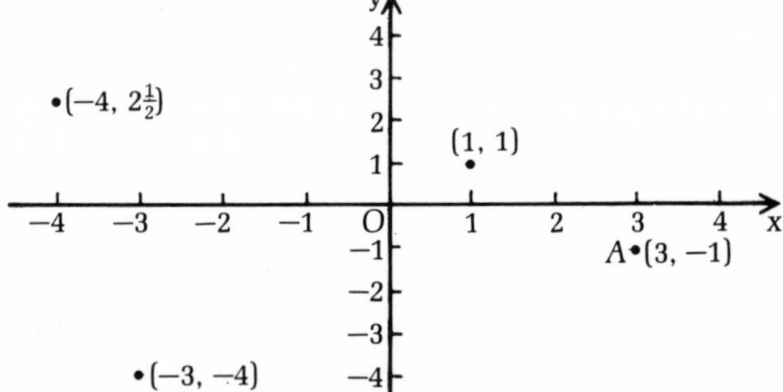

You should ensure that you can plot points.

Example: For point A (3, −1), go along the x-axis to 3 and down vertically to −1.

Interpretation and use of graphs

The following graph shows the comparison on a particular day of the values of pounds sterling (£) and US dollars ($).

The conversion graph enables us to change dollars into pounds or pounds into dollars.

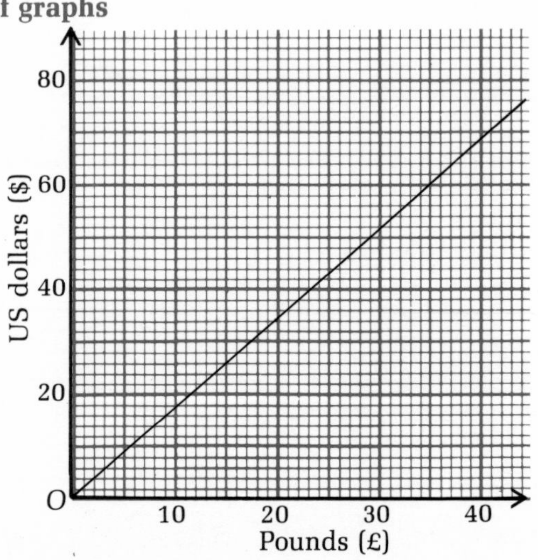

Examples 1: £20 is equivalent to 34 dollars (go along the £-axis until you reach 20 and then vertically until you reach the straight line – read this height off the vertical scale).

Likewise, in reverse 20 dollars is approximately £11·75.

Using the graph in this way is called *interpolation*.

Suppose you were asked to find what £100 is worth in dollars. You could extend the graph; or since you know £20 is equivalent to \$34, then £100 = 5 × £20 is equivalent to 5 × \$34, *i.e.* \$170. This process is called *extrapolation*.

In some cases you could be given information to represent graphically, or you might be given a time-distance graph to interpret.

Example 2: A motorist left Huddersfield at noon (1200 hours) and travelled to Scarborough, 136 km away. The graph of his journey is as shown. He had to stop for petrol and later had another stop for lunch.

From the graph determine the following.

a How far did he travel in the first hour?
b How long did his petrol stop take?
c What was the distance travelled between his two stops?
d How long did he take for the lunch stop?
e At what time did he reach Scarborough?
f At which stage of the journey was he travelling the fastest?
g How far was he from Scarborough at 1500 hours?

a 60 km *b* 10 min *c* 20 km *d* 40 min
e 1550 hours *f* the first stage *g* 32 km

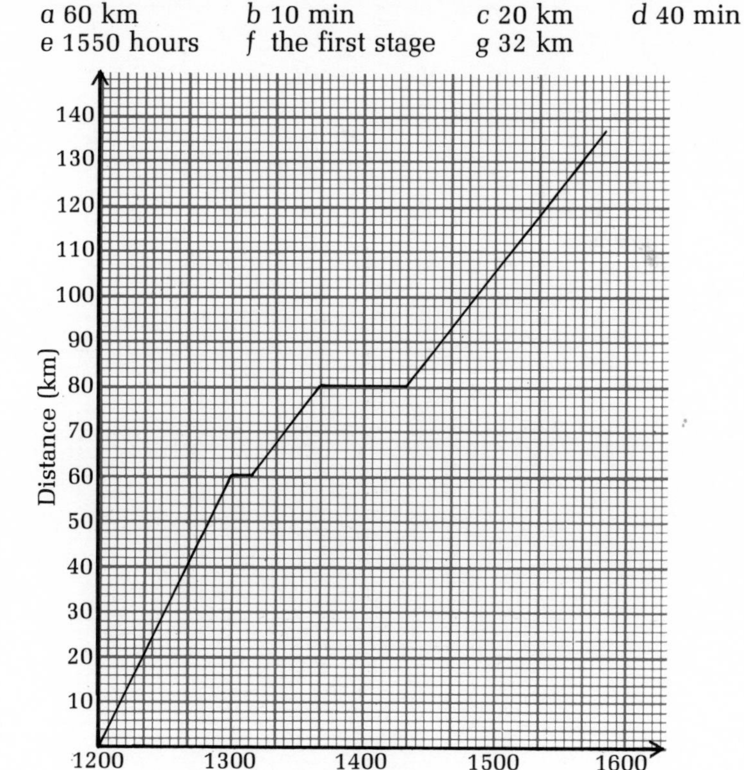

Intermediate Level

Plotting graphs of algebraic functions

At this level, we are mainly concerned with plotting graphs of certain algebraic functions, particularly of the forms

$$y = ax + b, \quad y = ax^2, \quad y = ax^2 + bx + c \quad \textit{and} \quad y = \frac{a}{x}$$

where a, b, c are integral constants.

In such cases we are normally given a range of values of x in which to plot the graph. x is called the *independent variable* and it is conventional to plot x horizontally. The value of y in each case will depend upon the value of x that is used, and hence y is called the *dependent variable*. The values of y are plotted vertically. It is more convenient to construct a table of values as shown below.

Example 1: Plot the graph of $y = 3x + 2$ for value of x from -2 to $+3$ inclusive. (This range might be shown as $-2 \leqslant x \leqslant 3$.)

Firstly make a table.

x	−2	−1	0	1	2	3
y	−4	−1	2	5	8	11

Now you can proceed.

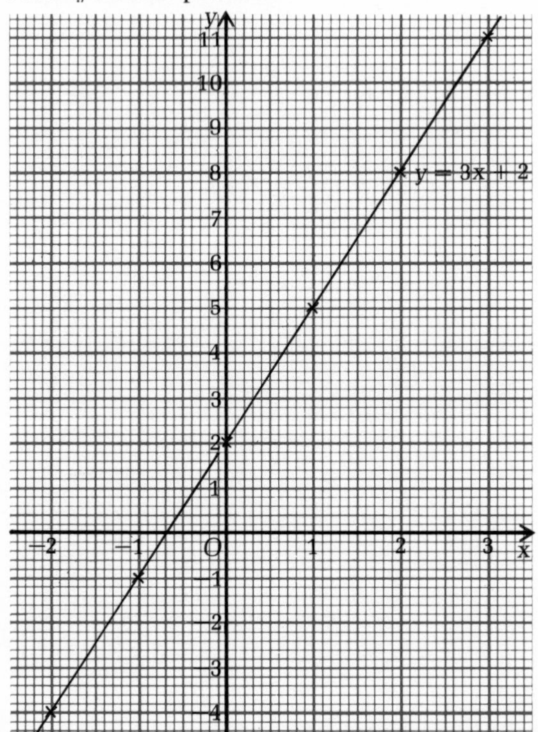

NOTE: For any equation of the form $y = ax + b$ where a and b are integral constants, the graph of y will always be a straight line and you need only two points to plot a straight line. Therefore if you recognise that the graph will be a straight line, there is no need to plot a table of values. If you are unsure, however, it might well be safer to plot the table.

Points are plotted by using either a small dot or a small cross. Use a ruler to draw the straight line.

Example 2: Draw the graph of $y = x^2 - 2x - 1$ for values of x in the range $-3 \leqslant x \leqslant 4$.

Make a table.

x	−3	−2	−1	0	1	2	3	4
x^2	9	4	1	0	1	4	9	16
$-2x$	6	4	2	0	−2	−4	−6	−8
−1	−1	−1	−1	−1	−1	−1	−1	−1
y	14	7	2	−1	−2	−1	2	7

Now you can proceed.

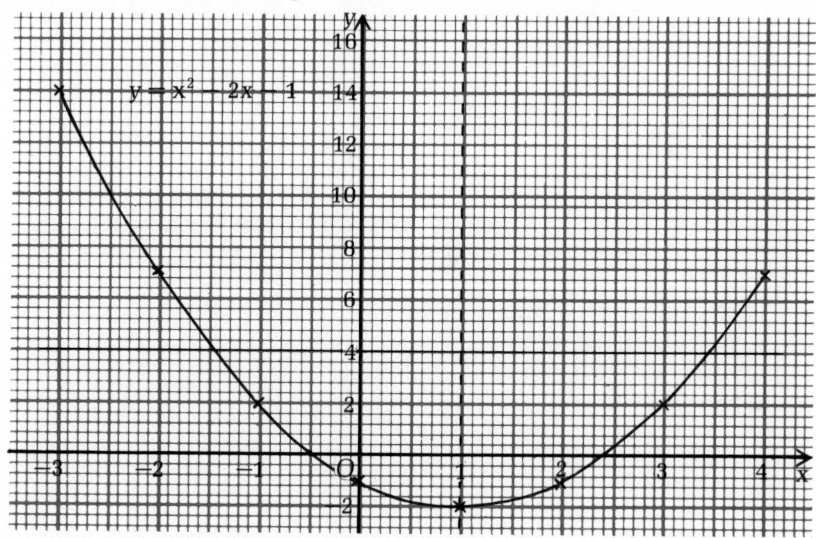

NOTE: Having plotted the points, connect them by a smooth freehand curve. You would lose marks for using a ruler and for a 'feathered' line.

The graph of $y = ax^2 + bx + c$ where a, b, c, are integers (except $a \neq 0$) is called a *parabola* and will always have symmetry about a vertical axis. The line $x = 1$ is the *axis of symmetry* for the above graph and is shown on the graph by the dotted line.

If a is positive, the graph will have the typical ∪ shape.

If a is negative, the graph will have the typical ∩ shape.

Example 3: Draw the graph of $y = \frac{2}{x}$ for values of $x > 0$ but $\leqslant 6$.

In cases where x appears in the denominator, you need to be careful as values of x approach zero. Therefore in this case consider some fractional values of x between 0 and 1.

x	$\frac{1}{4}$	$\frac{1}{2}$	$\frac{3}{4}$	1	2	3	4	5	6
y	8	4	$2\frac{2}{3}$	2	1	$\frac{2}{3}$	$\frac{1}{2}$	$\frac{2}{5}$	$\frac{1}{3}$

The graph of $y = \frac{a}{x}$, where x is some integer, is called a *hyperbola*.

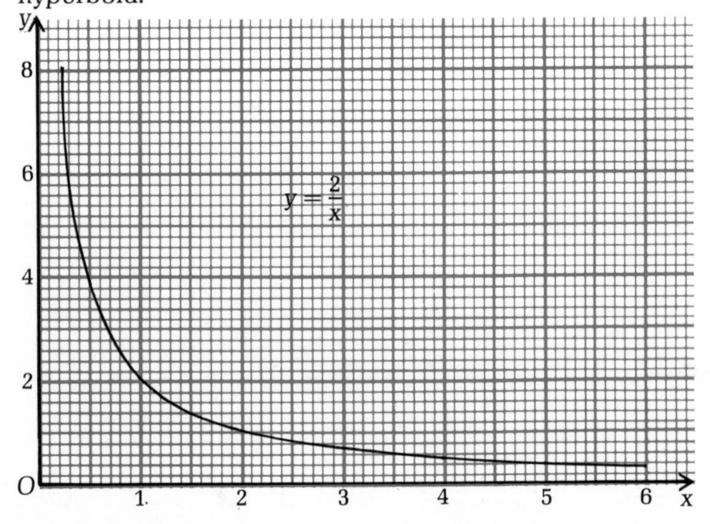

NOTE: For negative values of x, we obtain negative values of the above y values and the graph looks like

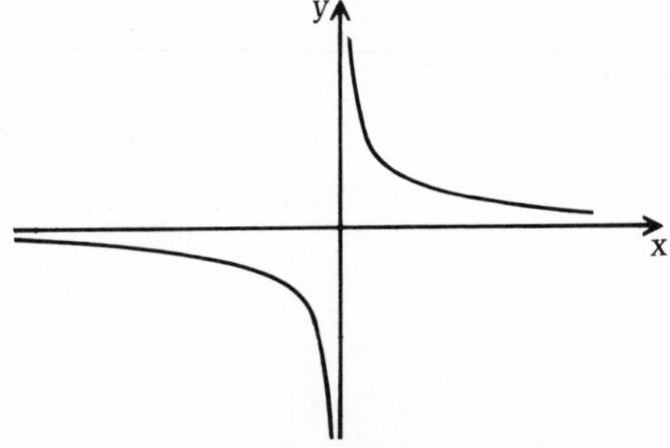

The points again should be connected by a smooth freehand curve.

Having drawn the graphs, you may be asked to find

a values from the graph **b** solutions to equations.

In *example 2:* What is the value of y when $x = 3\frac{1}{2}$?

Go along the x-axis until you reach $3\frac{1}{2}$, then go vertically up until you reach the curve and read off the value on the vertical scale. The exact value is $4\frac{1}{4}$, but you should be able to estimate some answer quite close to this value.

Again in *example 2:* Solve $x^2 - 2x - 1 = 0$.

We plotted the graph of $y = x^2 - 2x - 1$. If we take $y = 0$, we can read the solutions for x off the graph, i.e. where the curve cuts the horizontal axis.

The two values thus obtained are $x = -0.4$ and $x = 2.4$ approximately.

From example 2: Solve $x^2 - 2x - 5 = 0$. ①

We plotted $y = x^2 - 2x - 1$ ②

① can be rewritten $x^2 - 2x - 1 - 4 = 0$

$\therefore x^2 - 2x - 1 = 4$ ③

$\therefore$ equating ② and ③ $y = 4$

Now locate the line $y = 4$ on the graph. Where the line intersects the curve, read off the values of x to solve the equation $x = -1.45$ or 3.45 approximately.

Idea of gradient

The gradient of a straight line is a measure of its steepness.

Lines which slope upwards from left to right have a *positive* gradient.

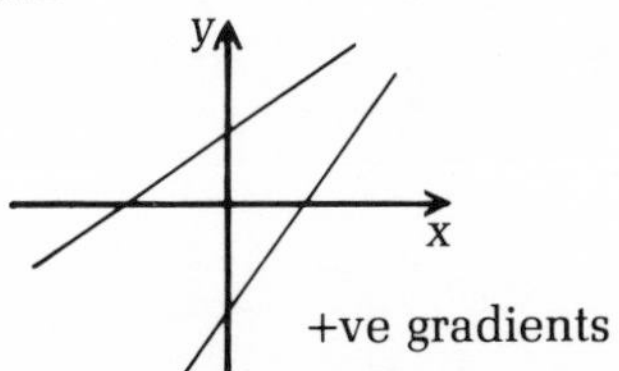

Lines which slope downwards from left to right have a *negative* gradient.

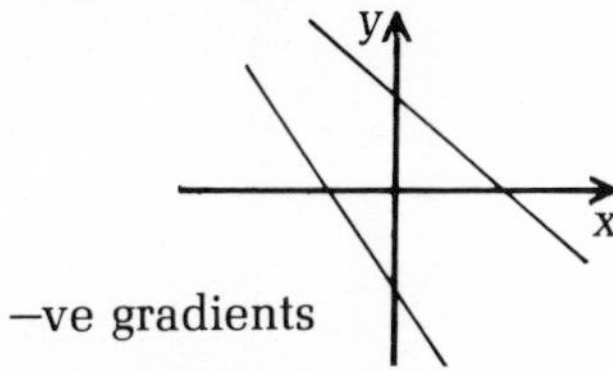

Horizontal lines have *zero* gradient (no slope).

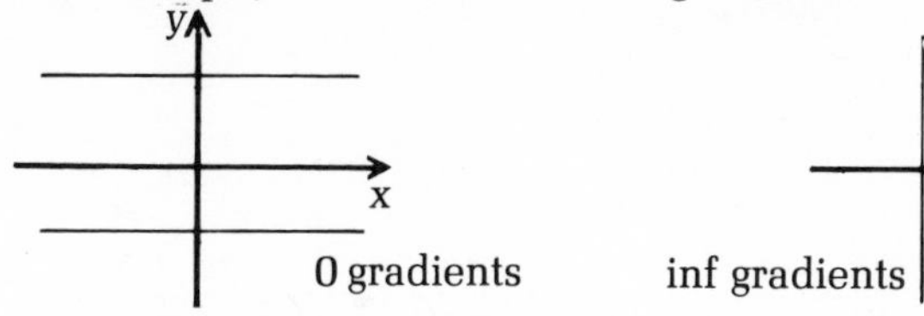

Vertical lines have an *infinite* gradient.

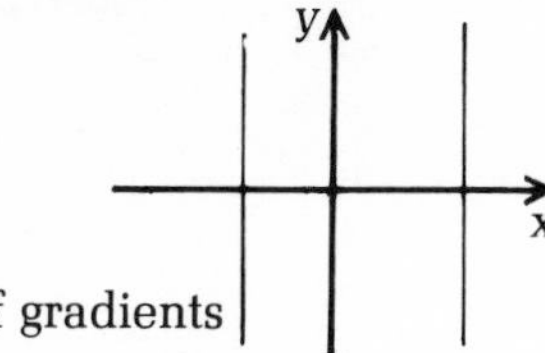

If we have two points (x_1, y_1) and (x_2, y_2) on a straight line, the gradient is given by $\frac{\text{change in } y}{\text{change in } x}$, *i.e.* $\frac{y_2 - y_1}{x_2 - x_1}$.

Example: Find the gradients of the lines joining the points given.

a (3, 5) (6, 9) gradient $= \frac{9-5}{6-3} = \frac{4}{3} = 1\frac{1}{3}$

b (7, 4) (8, 2) gradient $= \frac{2-4}{8-7} = \frac{-2}{1} = -2$

c (−2, 1) (4, 11) gradient $= \frac{11-1}{4--2} = \frac{10}{4+2} = \frac{10}{6} = 1\frac{2}{3}$

In a distance-time graph, the gradient of a straight line is the speed. In the earlier example for a distance-time graph, the speed for the first hour is

$$\frac{60-0}{13.00-12.00} = \frac{60}{1} = 60 \text{ km/h}.$$

Higher Level

Graphs of harder algebraic functions

The algebraic functions can be of the form $y = ax^3 + bx^2 + cx + d$, where usually at least one of a, b, c and d is zero, or $y = \frac{k}{x^2}$ $(x \neq 0)$.

Example 1: Plot $y = x^3 + 2x$ for values of x in the range $-3 \leqslant x \leqslant 3$.

x	−3	−2	−1	0	1	2	3
x^3	−27	−8	−1	0	1	8	27
2x	−6	−4	−2	0	2	4	6
y	−33	−12	−3	0	3	12	33

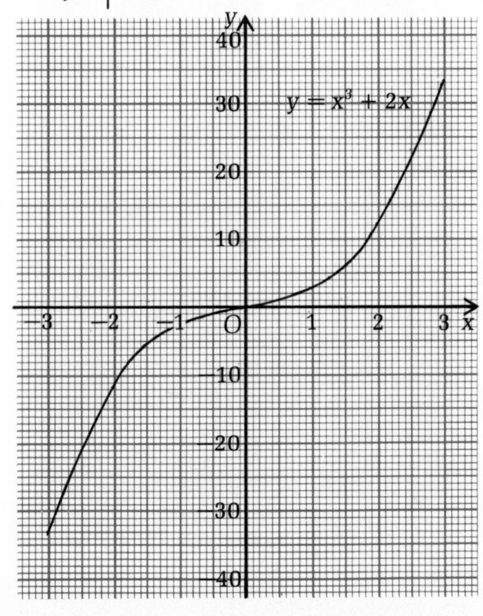

Values might be found or equations solved as mentioned earlier.
The equation $y = ax + b$ always represents the graph of a straight line. Often a straight line is represented by $y = mx + c$; in this case we have simply changed the letters, m for a, c for b.

Example 2: Draw the graph of $y = 3x + 1$ for values of x in the range $x \geqslant -1$ and $x \leqslant 4$.
The graph will be a straight line, so only two points are needed.
When $x = -1$, $y = 3(-1) + 1 = -3 + 1 = -2$.
When $x = 4$, $y = 3(4) + 1 = 12 + 1 = 13$.

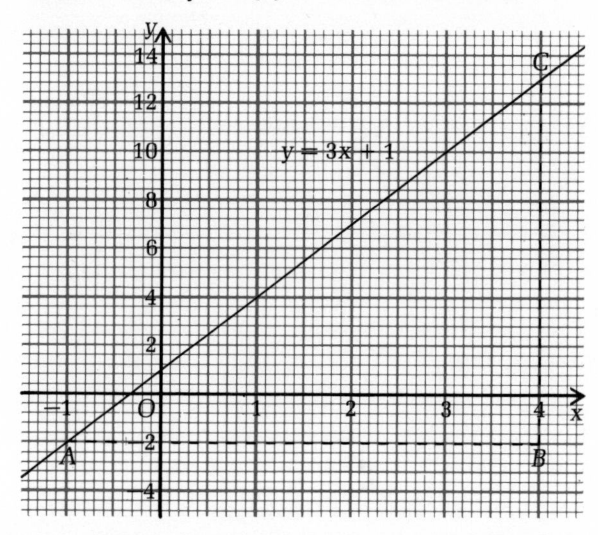

NOTE: The line cuts the vertical axis at $y = 1$. For $y = 3x + 1$, comparing to $y = mx + c$ gives $c = 1$.

$\therefore$ the intercept on the y-axis is c.

Two points on the line have been calculated $(-1, -2)$ and $(4, 13)$.

The gradient of the line is $\dfrac{y_2 - y_1}{x_2 - x_1} = \dfrac{13 - -2}{4 - -1} = \dfrac{13 + 2}{4 + 1} = \dfrac{15}{5} = 3$

or on the diagram, gradient $= \dfrac{CB}{AB} = \dfrac{15}{5} = 3$.

Comparing $y = 3x + 1$ and $y = mx + c$, $3 = m =$ gradient.

For $y = mx + c$, the gradient is always m, the intercept on y-axis is always c.

Further examples 1: If $y = -3x + 7$, gradient is -3, intercept on y-axis $= 7$

2: If $y = 2x - 9$, gradient is 2, intercept on y-axis $= -9$

NOTE: *3*: If $2y = 7x + 4$, we must first divide throughout by 2 to compare with $y = mx + c$,

i.e. $y = \frac{7}{2}x + \frac{4}{2}$ $\therefore y = 3.5x + 2$

so gradient is 3.5, intercept on y-axis is 2.

Inequalities

Example 1: If $3x + 7 \geqslant 13$
then $3x \geqslant 13 - 7$
$3x \geqslant 6$
$x \geqslant \frac{6}{3}$, i.e. $x \geqslant 2$

Example 2: If $2x \leqslant x + 3$
$2x - x \leqslant 3$
$x \leqslant 3$

Inequalities can be represented graphically.

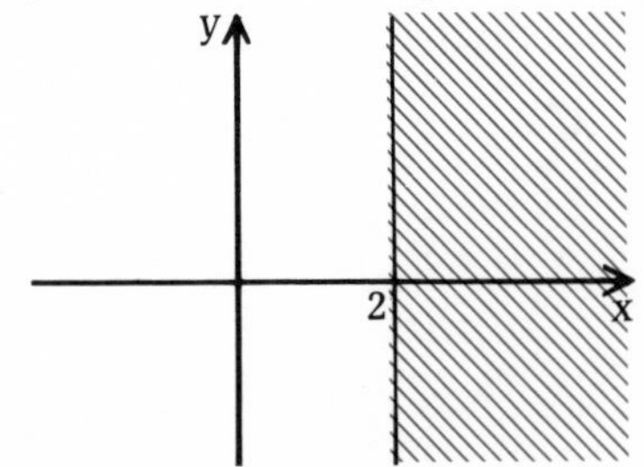

(i) The shaded portion is $x \geqslant 2$.

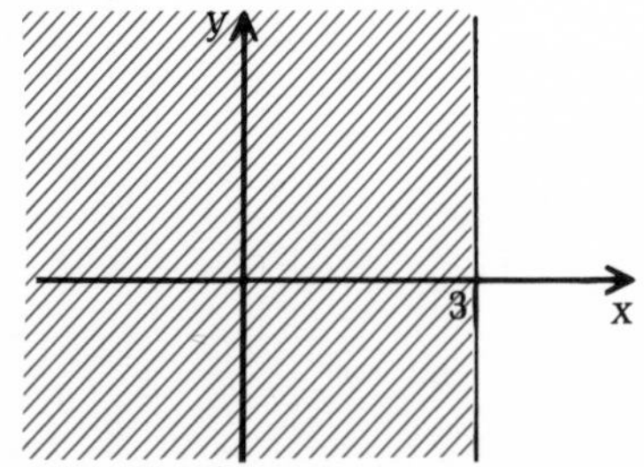

(ii) The shaded portion is $x \leqslant 3$.

Suppose we are given the following set of inequalities: $y \leqslant 3x$, $y \leqslant 2 - x$, $y \geqslant -3$. Indicate the region **A** which is satisfied by these inequalities on a graph, by considering values of x in the range $-2 \leqslant x \leqslant 6$.

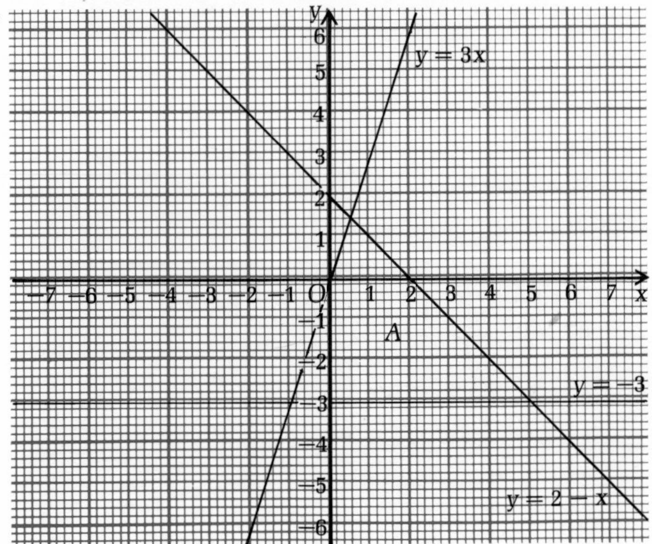

In such problems we shall be concerned only with the graphs of straight lines. Therefore plot the graphs of $y = 3x$, $y = 2 - x$, and $y = -3$.

The region **A** can be found in two different ways, either by shading 'in' or shading 'out'. If we shade 'in', we shade the given requirements and area **A** is what will be treble shaded.

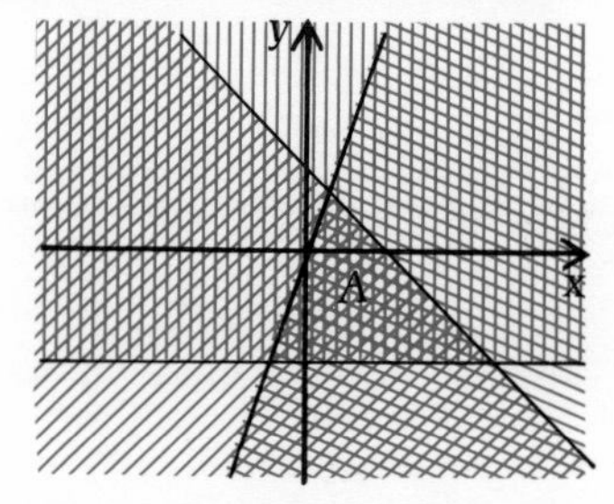

$y \geqslant -3$
$y \leqslant 3x$
$y \leqslant 2 - x$

If we shade 'out', we shade the areas not required.

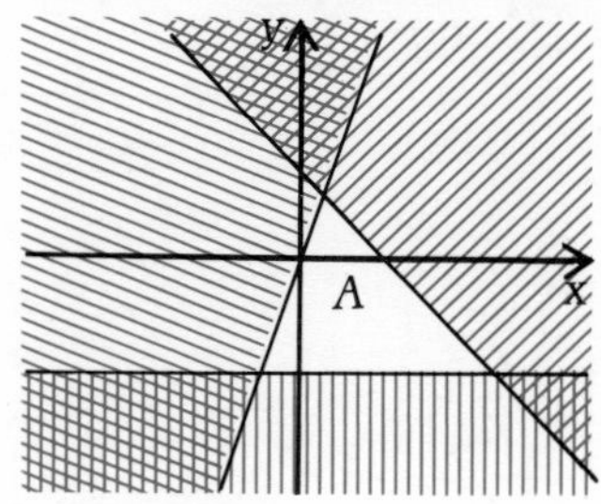

$y < -3$
$y > 3x$
$y > 2 - x$

A is the unshaded area.

13 Geometric Terms

Foundation Level
Angles, descriptions of acute, right, obtuse, straight line and reflex angles, more about angles, similarity.

Intermediate Level
Congruence.

Foundation Level

Angles

An angle is formed when two straight lines meet at a point. For instance *BA* and *BC* meet at *B* to form an angle. The angle is denoted by the shorthand $\angle B$ or $\angle ABC$. In the first case, *B* is the point where the lines meet; the second describes the meeting of *BA* and *BC* which is more explicit.

The size of the angle depends upon the amount of turn between *BC* and *BA* and is independent of the length of the lines *BC* and *BA*. Sometimes, a Greek letter is used, e.g. θ.

If *BC* is rotated about *B* one complete turn in an anticlockwise direction, it is said to have turned through 360 degrees which is denoted by 360°.

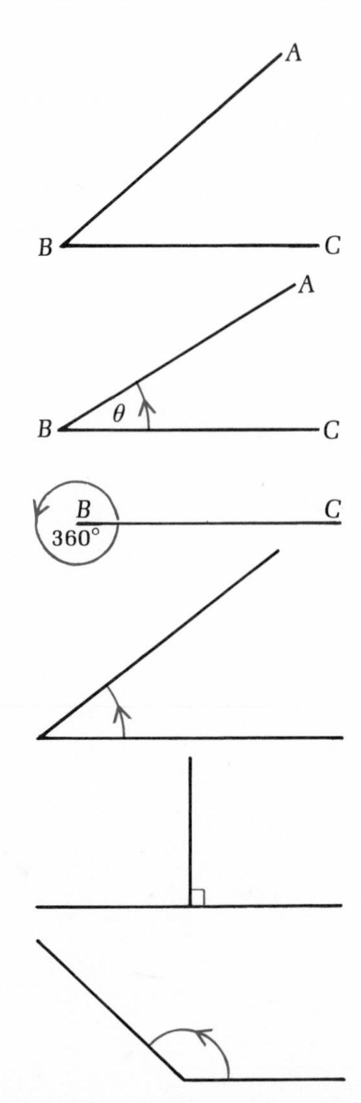

Acute angle

An acute angle is an angle between 0° and 90°.

Right angle

A right angle is an angle of 90° denoted as shown.

Obtuse angle

An obtuse angle is an angle between 90° and 180°.

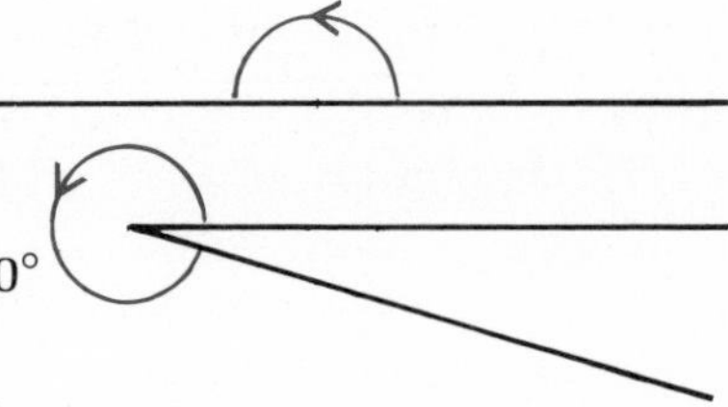

Straight line angle

A straight line angle is 180°.

Reflex angle

A reflex angle is an angle between 180° and 360°.

More about angles

Two angles which when added together make 90° are said to be *complementary*. The complementary angle of 60° is therefore 30°.

Two angles which together make 180° are said to be *supplementary*. The supplementary angle of 80° is therefore 100°.

When a right angle is formed between two lines, the lines are said to be *perpendicular* to each other.

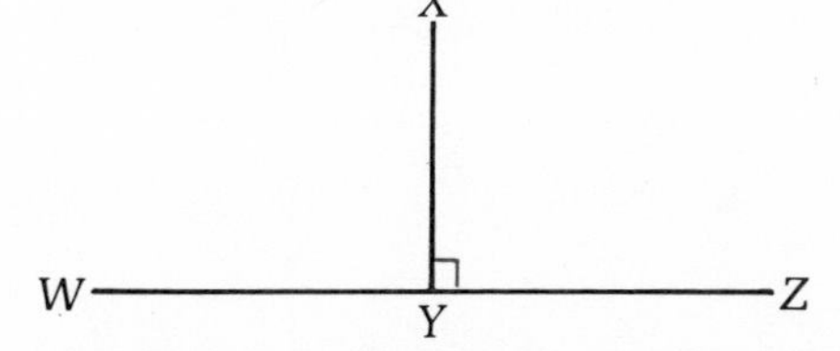

XY is perpendicular to *WYZ*.

When two straight lines intersect, the opposite angles so formed are equal and are called *vertically opposite* angles.

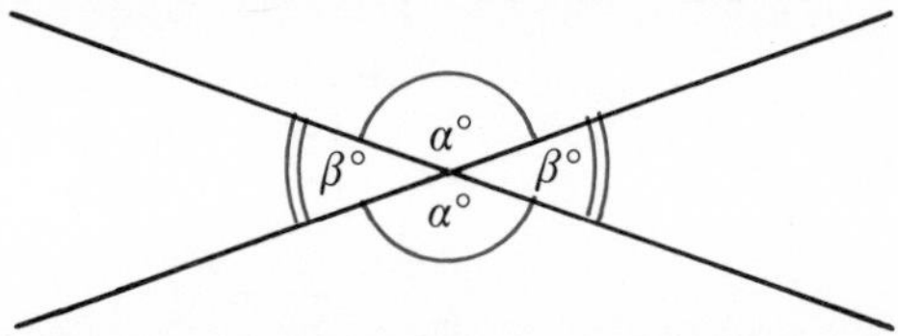

Parallel lines never meet and they are usually marked with arrows to indicate that they are parallel. In the diagram, *AB* and *CD* are parallel.

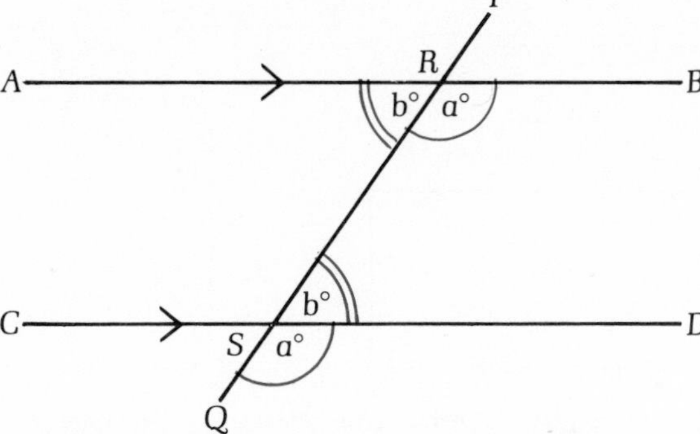

If a straight line *PQ* cuts the parallel lines at *R* and *S* as shown, *PQ* is called a *transversal* and some very important angle properties occur.

Corresponding angles are equal, marked $a°$, sometimes called F angles.
Alternate angles are equal, marked $b°$, sometimes called Z angles.
NOTE: At point R, $a° + b° = 180°$ (angles on a straight line)
$\therefore\ \angle BRS + \angle DSR = 180°$
i.e. interior angles between parallel lines sum to 180°.
There are other alternate, corresponding and vertically opposite angles that can be shown on the diagram.

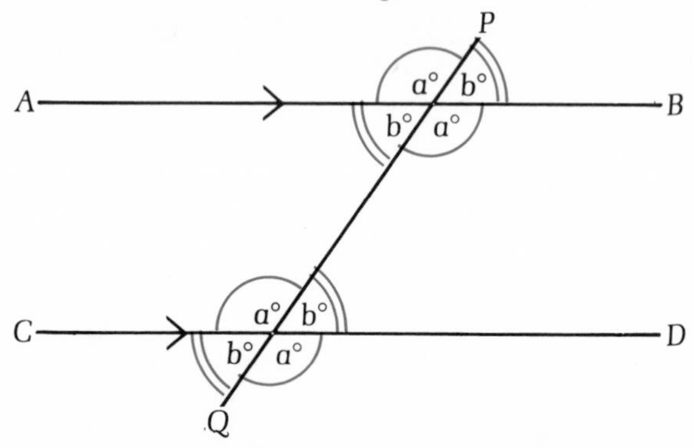

Similarity

When two figures have angles which correspond to each other and the corresponding lengths of the sides are in the same ratio to each other, the figures are said to be similar.

Example:

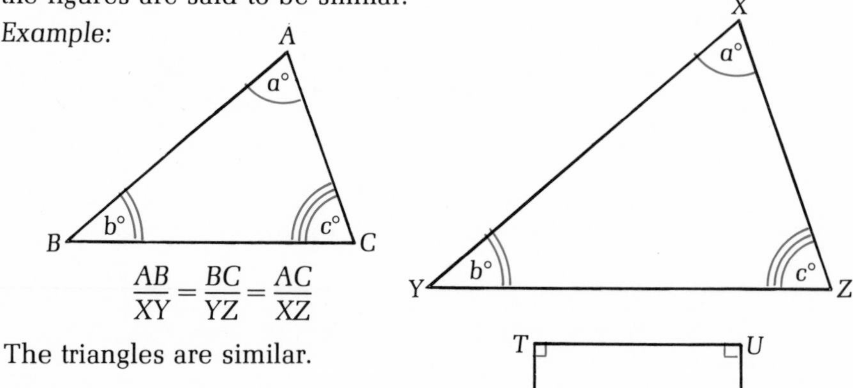

$$\frac{AB}{XY} = \frac{BC}{YZ} = \frac{AC}{XZ}$$

The triangles are similar.

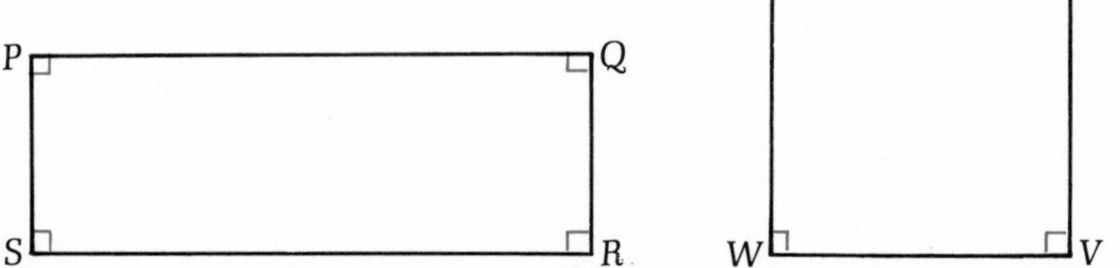

Although the angles correspond, the ratio of corresponding sides is not the same. Therefore these rectangles are not similar.

Intermediate Level

Congruence

Figures which have the same size and shape are said to be congruent, *i.e.* they could be superimposed on each other.

We are generally more concerned with congruent triangles and there are four types to be considered. It helps to mark equal angles and equal lengths of sides as shown.

Type A

One side and two angles of one triangle equal to one side and two angles of a second triangle.

Of course, in the above, the third unmarked angles will also be equal. This is often called AAS – for angle, angle, side.

Type B

Two sides and the angle between them in one triangle equal to two sides and the angle between them of a second triangle.

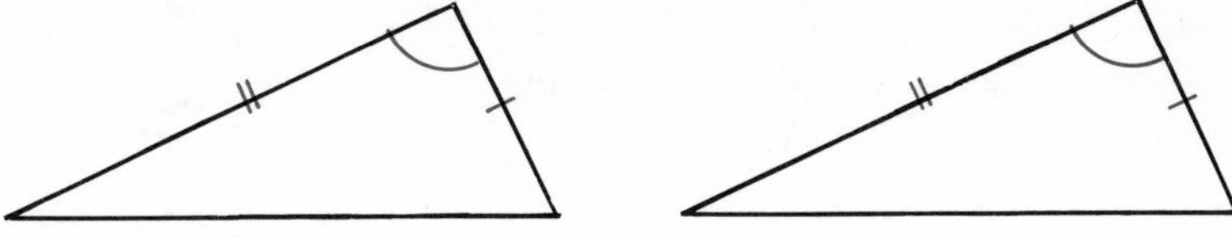

This is often called SAS – for side, angle, side.

Type C

Three sides of one triangle equal to three sides of a second triangle.

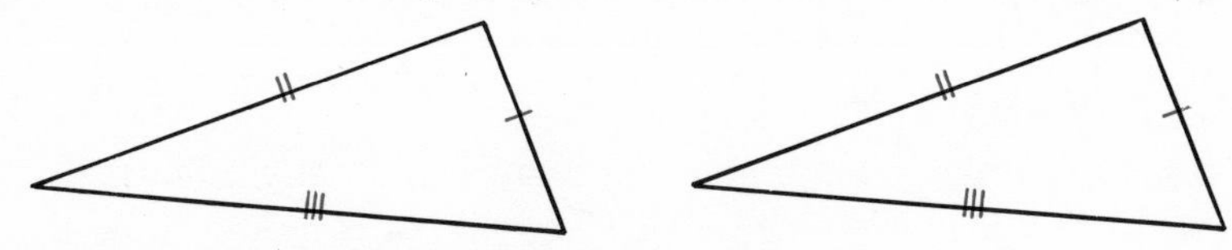

This is often called SSS for side, side, side.

Type D

In right-angled triangles, the side opposite the right angle (called the *hypotenuse)* and one other side in each triangle equal.

This is often called RHS – for right angle, hypotenuse and side.

14 Polygons and Circles

Foundation Level

Polygons, description of different types of triangle, description of different types of quadrilateral, interior angle sum of triangle and quadrilateral, circles.

Intermediate Level

Angle properties of a regular polygon (including sum of exterior angles), angle properties of a circle.

Higher Level

Interior angle sum of polygons.

Foundation Level

Polygons

A polygon is the general name given to two-dimensional shapes with three or more straight sides. The two simplest polygons are *triangles* (three sides) and *quadrilaterals* (four sides.

We shall now consider the different types of each.

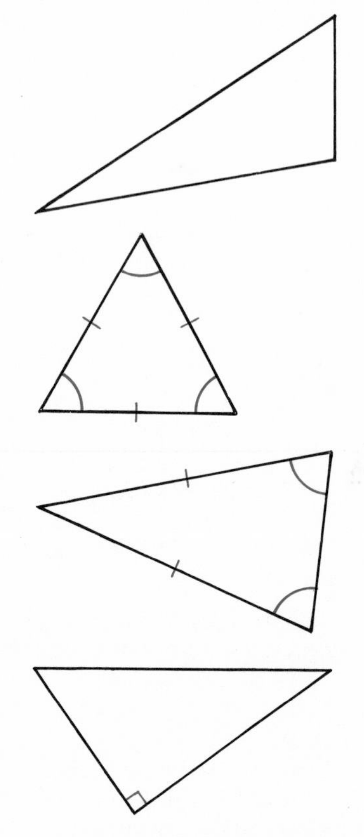

Scalene triangle

A triangle with all sides and angles different is called a scalene triangle.

Equilateral triangle

A triangle with three equal sides and three equal angles is called an equilateral triangle.

Isosceles triangle

A triangle with two sides equal and two angles equal is called an isosceles triangle.

NOTE: In such a case, the unequal angle is always the angle *between* the equal sides.

Right-angled triangle

A triangle where one of the angles is a right angle is said to be a right-angled triangle. The side opposite the right angle is called the *hypotenuse*. The right angle is marked as shown.

Sum of the interior angles of a triangle

If any triangle is drawn on paper, with the angles marked, and then torn into three pieces as shown, it can be demonstrated that the three angles placed together form an angle on a straight line. Thus the interior angles of a triangle always add up to 180°.

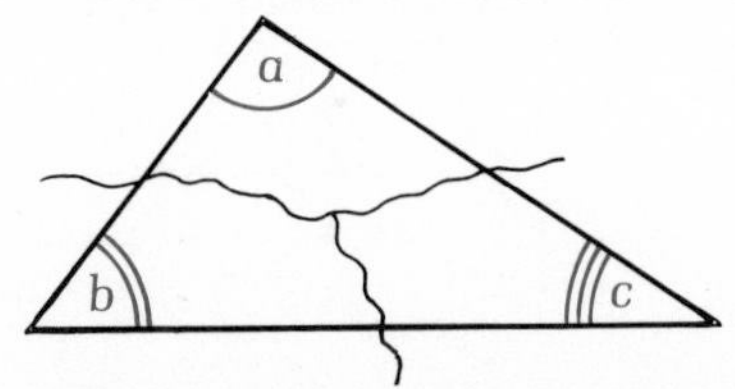

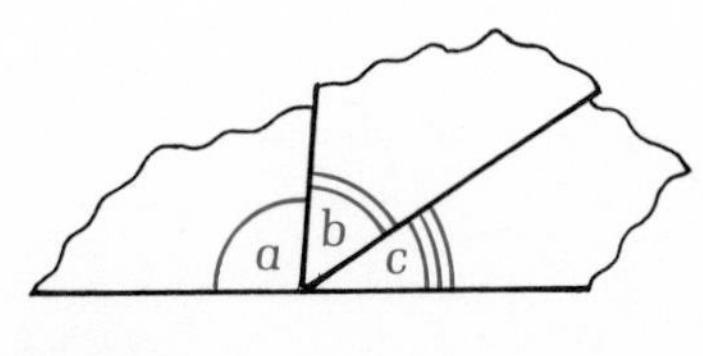

Sum of the interior angles of a quadrilateral

By drawing any quadrilateral and marking the angles, then tearing the quadrilateral into four pieces as shown, it can be demonstrated that the four angles placed together form an angle of 360°. Thus the interior angles of any quadrilateral always add up to 360°.

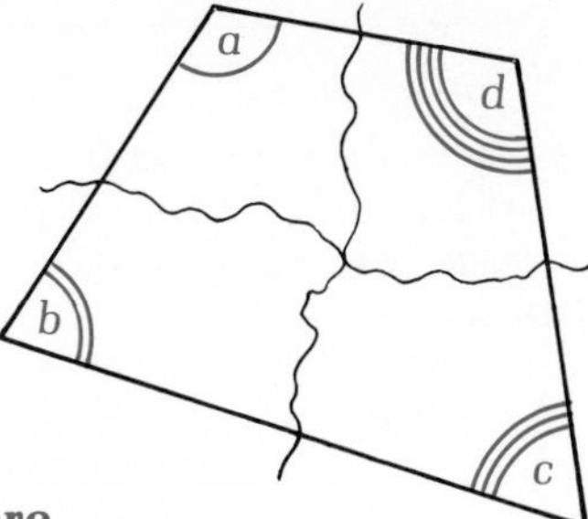

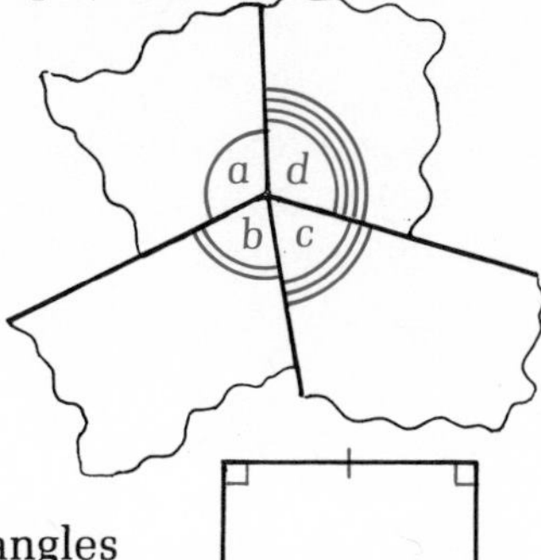

Square

This is a quadrilateral with four right angles and four sides of equal length.

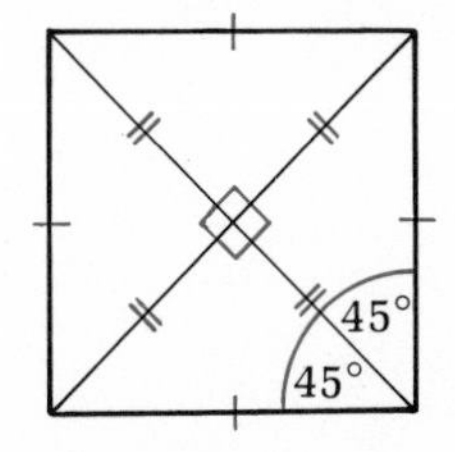

A *diagonal* is a straight line joining opposite corners of a quadrilateral. The two diagonals of a square are equal in length, they bisect each other at right angles and they bisect the right angles.

Rectangle

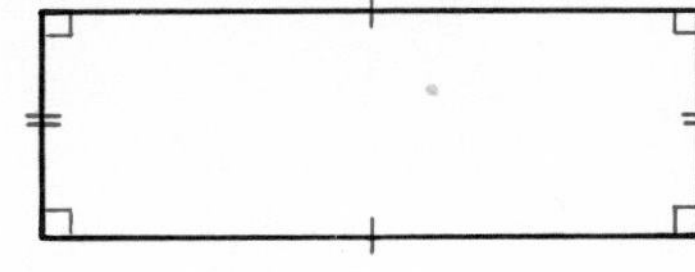

This is a quadrilateral with four right angles, opposite sides equal in length but adjacent sides of different length.

The two diagonals of a rectangle are equal in length and bisect each other. Note that they do not bisect the right angles.

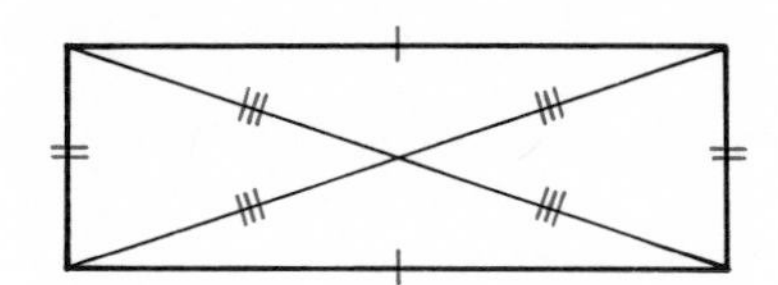

Parallelogram

This is a quadrilateral with both pairs of opposite sides parallel.

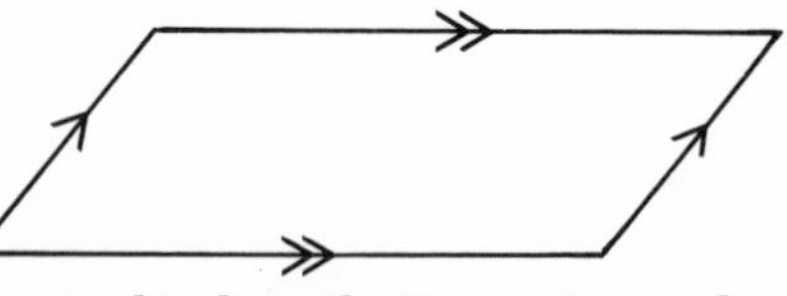

A parallelogram has its opposite sides equal in length. Opposite angles are equal. The diagonals bisect each other and divide the parallelogram into two congruent triangles.

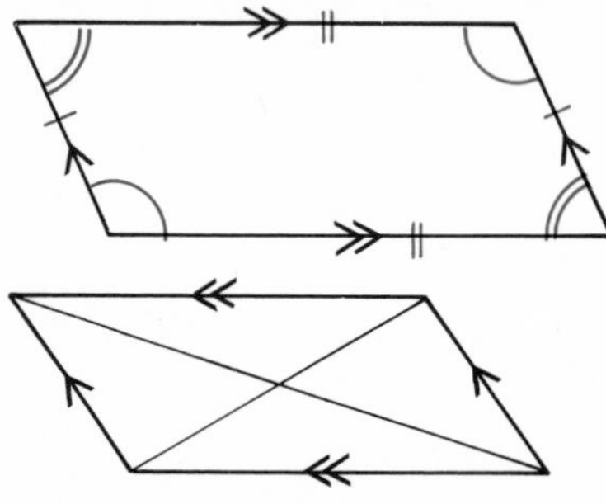

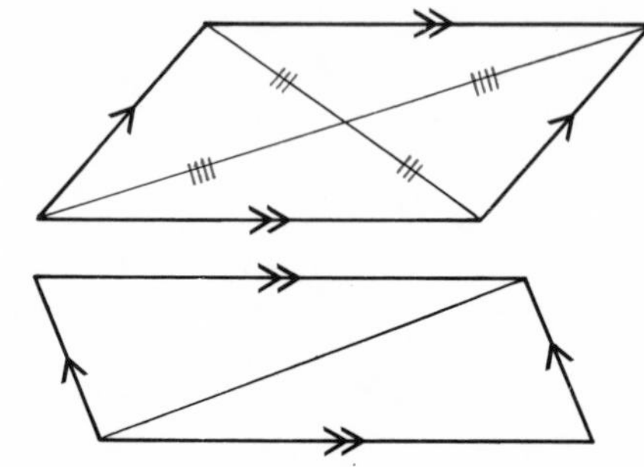

Rhombus

This is a quadrilateral with all sides equal, opposite sides parallel and opposite angles equal.

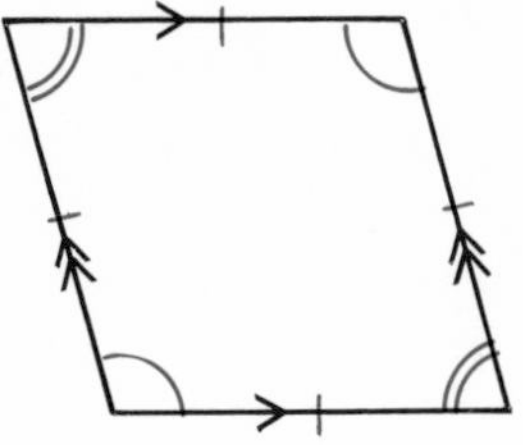

The diagonals of a rhombus bisect each other at right angles and bisect the angles through which they pass.

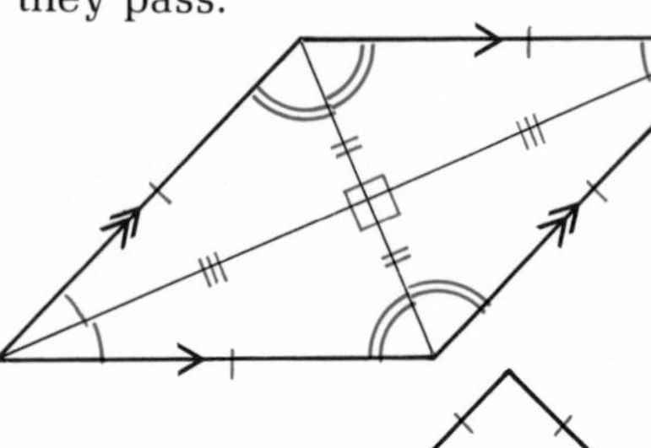

Kite

This is a quadrilateral with one pair of adjacent sides equal in length and the other pair of adjacent sides equal in length.

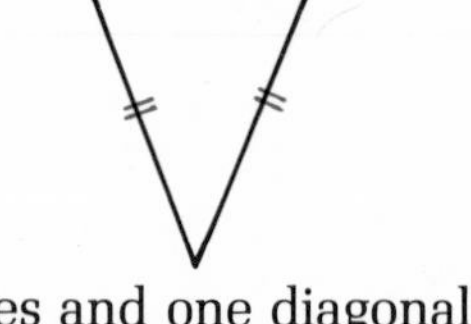

There is one pair of diagonally opposite equal angles and one diagonal forms two congruent triangles.

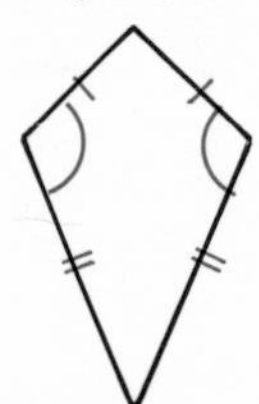

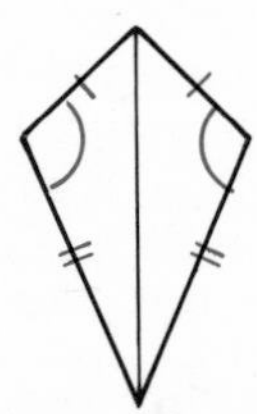

Trapezium

This is a quadrilateral with one pair of opposite sides parallel.

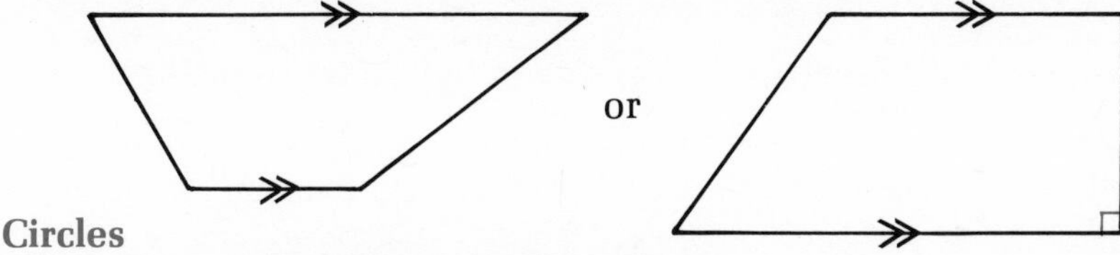

Circles

A circle is a plane figure bounded by a single curved line, called its *circumference*. Every point of the circumference is equally distant from a point within the figure, called the *centre*.

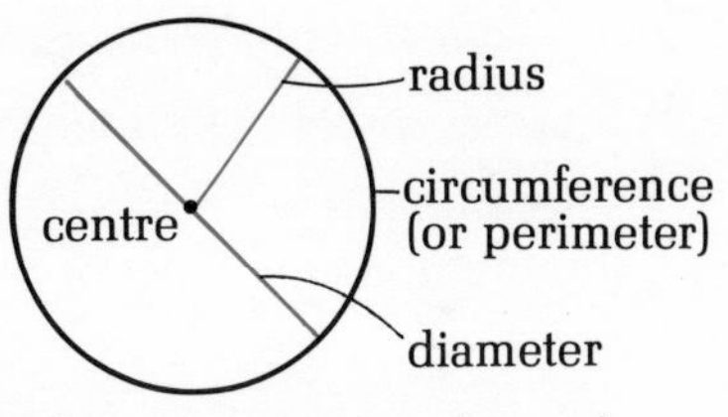

The distance from centre to circumference is called a *radius*.

Any line drawn from two points on the circumference passing through the centre of the circle is called a *diameter*.

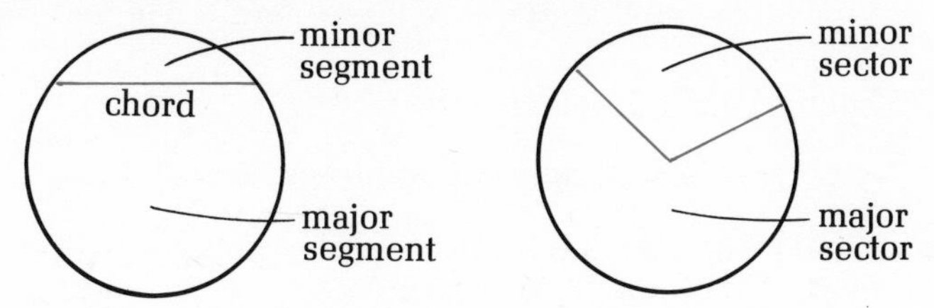

Any line joining two points on the circumference is called a *chord*. The chord divides the circles into two *segments* – a minor and a major segment.

Any two points on the circumference joined to the centre of a circle form two *sectors* – a minor and a major sector.

Intermediate Level

Angle properties of a regular polygon

At this level you are only concerned with regular polygons. A regular polygon is a polygon with equal side lengths, equal interior angles and hence equal exterior angles.

An equilateral triangle and a square are examples of regular polygons.

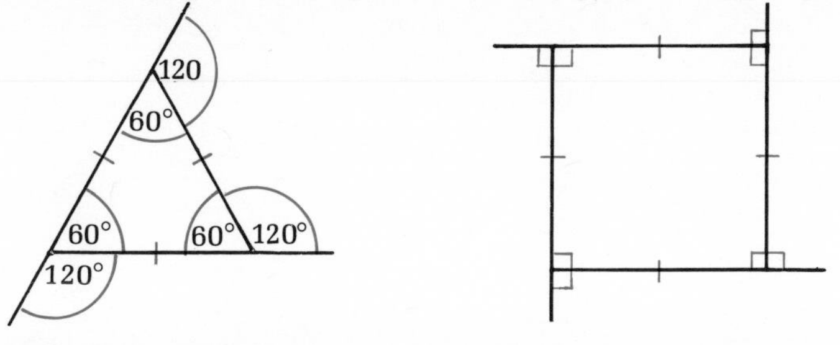

Exterior angles are formed by extending one side at each vertex of a polygon.

Exterior + Interior angles = 180° at any vertex.

NOTE: The sum of the exterior angles of any polygon is 360°.

A polygon has three or more sides. Some of the more common ones, other than a triangle and a quadrilateral, are as follows.

5 sides – pentagon	8 sides – octagon
6 sides – hexagon	9 sides – nonagon
7 sides – heptagon	10 sides – decagon

The questions on the above will always involve regular polygons and the fact that the sum of the exterior angles of a polygon is 360°.

Example 1: If a regular polygon has 12 sides, what is the size of each interior angle?

Sum of exterior angles = 360°.

Since polygon is regular, there are 12 equal exterior angles.

Each exterior angle $= \frac{360°}{12} = 30°$

$\therefore$ each interior angle = 180° − 30° = 150°

Example 2: If each interior angle of a regular polygon is 135°, how many sides has the polygon?

Each exterior angle = 180° − 135° = 45°

$\therefore$ number of exterior angles $= \frac{360°}{45°} = 8$

$\therefore$ there are 8 sides to the polygon (there are as many sides as exterior angles)

Angle properties of a circle

Although there are many angle properties of a circle, you are only concerned with two at this level.

The angle in a semicircle

If any point on the circumference of a circle is joined to the ends of a diameter of the circle, then the angle so formed is a right angle.

Since a diameter bisects a circle, *i.e.* it divides the circle exactly into 2 parts, called semicircles, the angle subtended by the diameter at the circumference is called an angle in a semicircle.

The angle in a semicircle is 90°.

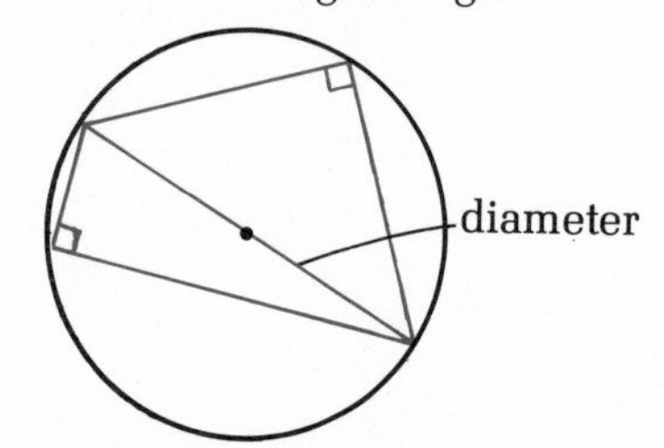

The angle between a tangent to a circle and the radius

A *tangent* to a circle is a straight line touching the circle at one point on the circumference only. This point is called the point of contact.

The angle between a tangent and the radius at the point of contact is always a right angle.

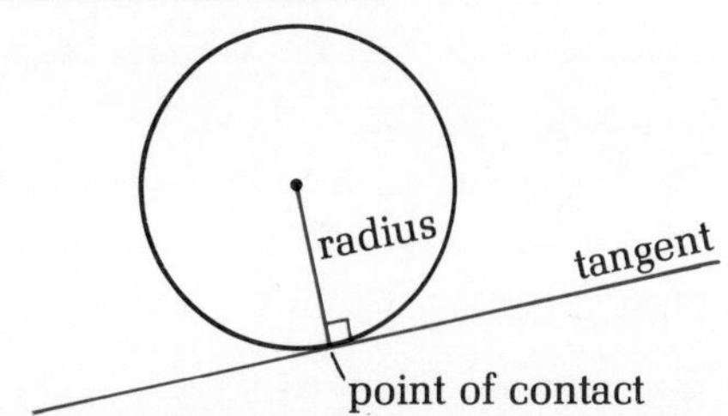

Higher Level

Sum of the interior angles of a polygon

The angle sum of the interior angles of a triangle is 180° or 2 right angles.

The angle sum of the interior angles of a quadrilateral is 360° or 4 right angles. This has already been shown.

We could also split the quadrilateral into two triangles. Hence, sum of interior angles is $2 \times 180° = 360°$ or 4 right angles.

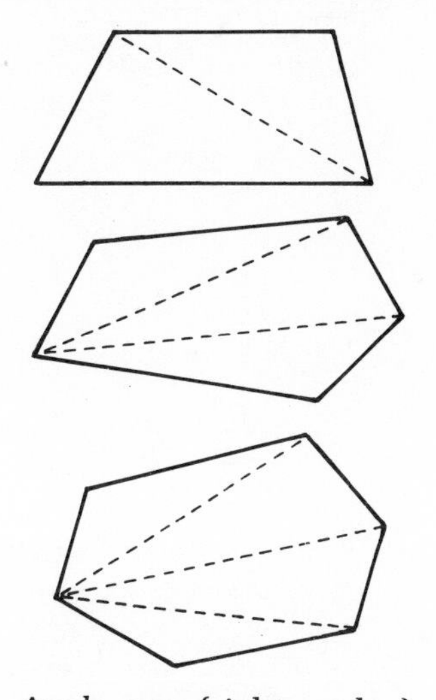

A pentagon can be split into 3 triangles. Hence, sum of interior angles is $3 \times 180° = 540°$ or 6 right angles.

A hexagon can be split into 4 triangles. Hence, sum of interior angles is $4 \times 180° = 720° = 8$ right angles.

Number of sides	*Number of triangles*	*Angle sum* (right angles)
3	1	2
4	2	4
5	3	6
6	4	8
n	$(n - 2)$	$2(n - 2)$

$\therefore$ sum of interior angles of an n-sided polygon is $(2n - 4)$ right angles or $90(2n - 4)° = (180n - 360)°$.

Example: If 4 angles of a pentagon are 136°, 81°, 90°, 118°, find the fifth angle.

Sum of angles of a pentagon $= (2 \times 5 - 4)$ right angles
$= 6$ right angles $= 540°$

Sum of four angles $= 136° + 81° + 90° + 118° = 425°$

$\therefore$ fifth angle $= 540 - 425 = 115°$

15 Drawing, Bearings and Scale

Foundation Level
Construction of a triangle, bearings, scale drawings.

Foundation Level

Construction of a triangle

It is essential that you have a ruler, a hard sharp pencil, a protractor and a pair of compasses. Marks will be lost in examinations for inaccurate measurements, thick unsightly lines, etc. The only ruler-and-compasses construction in the exam will be for a triangle with sides of given length. You will not be required to construct specific angles.

Example: Construct a triangle ABC in which $AB = 10$ cm, $AC = 8$ cm, $BC = 7$ cm.

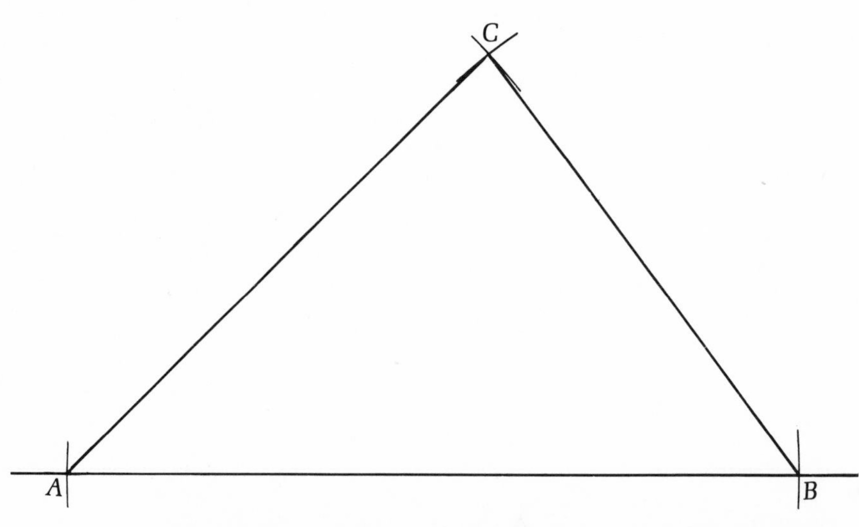

Remember to leave a reasonable amount of space before drawing a straight line. Measure as accurately as possible 10 cm with compasses, and place at A on line and make arc at B. Then with centre B, and same arc, mark at A. With centre A and radius 8 cm, make a large arc above AB. With centre B and radius 7 cm, make an arc to cut the previous arc to fix point C. Join C to A and B.

If you are asked to measure the angles, then use your protractor as accurately as possible.

$\angle A = 44°$ $\angle B = 53°$ $\angle C = 83°$ (to nearest degree)

Other types of figure may involve the use of a protractor.

Example: Construct a triangle XYZ so that angle X is 65°, XY is 8 cm, angle Y is 55°.

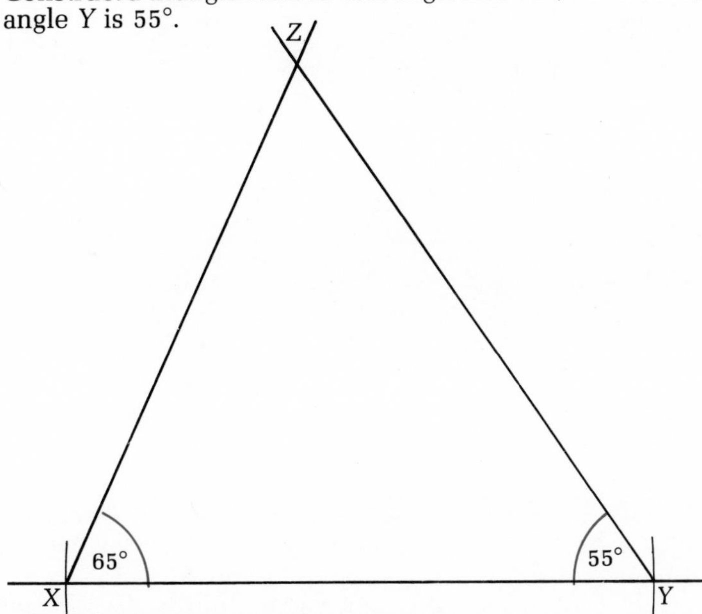

Draw a straight line. Using ruler and compasses measure $XY = 8$ cm. Use protractor to measure angle $X = 65°$ and draw the arm of the angle from X. Repeat for angle Y. Where the two arms meet fixes Z.

It is difficult to get exact measures, but you must aim to construct and measure as accurately as possible.

Bearings

The direction in which one point is fixed relative to some other point is given by a bearing. One method is by reference to the four cardinal directions, North, East, South and West.

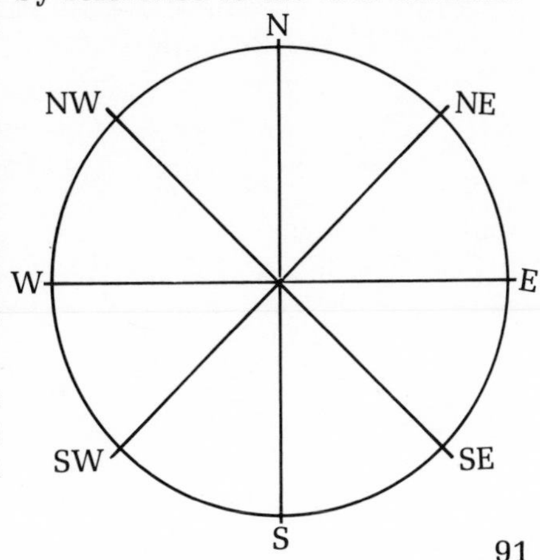

The eight directions of North, North-East, East, South-East, South, South-West, West and North-West are frequently used in everyday life, and these eight points of the compass are the only ones that will be used in the examination for this particular method.

A likely question will involve the use of three-figure bearings measured from a North line in a clockwise direction.

The following diagram shows bearings of A, 030°; B, 098°; C, 220°; D, 330° from the point O.

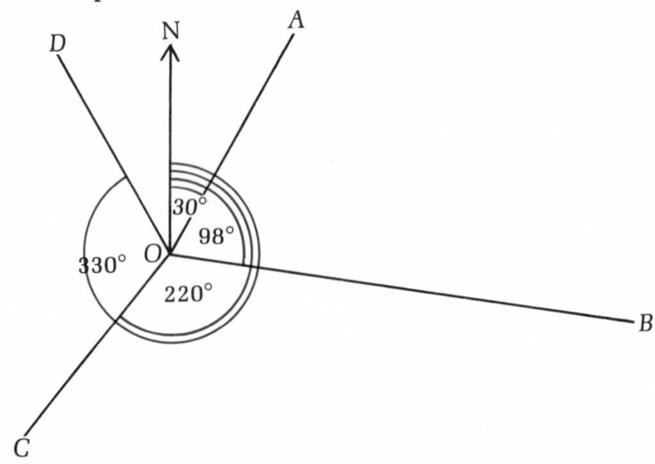

In such questions on bearings, the important word is *from*. Always draw a North line at the point where the bearing is measured *from*.

Example:

Find the bearing of
a Sheffield from Manchester
b Manchester from Sheffield.

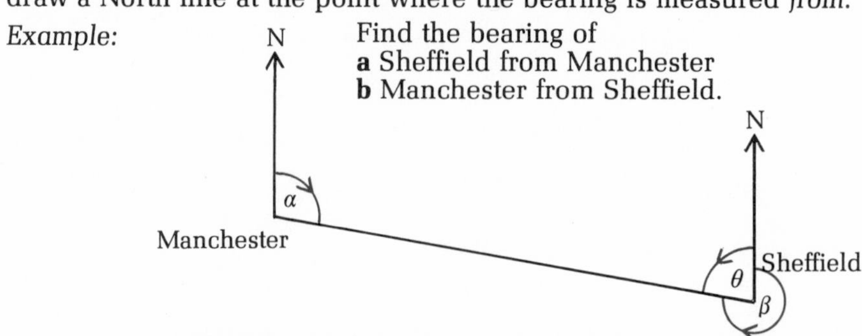

a From Manchester, draw a North line. Join Manchester to Sheffield. By measurement, $\alpha = 100°$.
$\therefore$ bearing is 100°.

b From Sheffield, draw a North line. To find β, measure θ, and take θ from 360° for the bearing. $\theta = 80°$.
$\therefore$ bearing is $360° - 80° = 280°$.

Scale drawings

In everyday life, we use maps of countries, street plans of towns, pictures of various objects, house plans, etc. which are clearly representative and drawn to scale to depict whatever is under consideration. Scales can be expressed in two ways.

Type A

An actual length of 10 km could be represented on a diagram by, say, 10 cm so that the scale is 1 cm represents 1 km.

Type B

Since 1 km = 100 000 cm, the scale ratio is 1:100 000. No units are involved if the scale is given in this form of 1:n where n is called the scale factor. In the example the scale factor is 100 000.

Depending upon what is under consideration, we can scale up or down to obtain enlargements or reductions, respectively. In the examination, you could be asked to interpret scale drawings or to make a scale drawing of some situation.

Example 1: This plan of a rectangular lawn is drawn to a scale 2 mm to 1 m. What are the actual dimensions of the lawn?

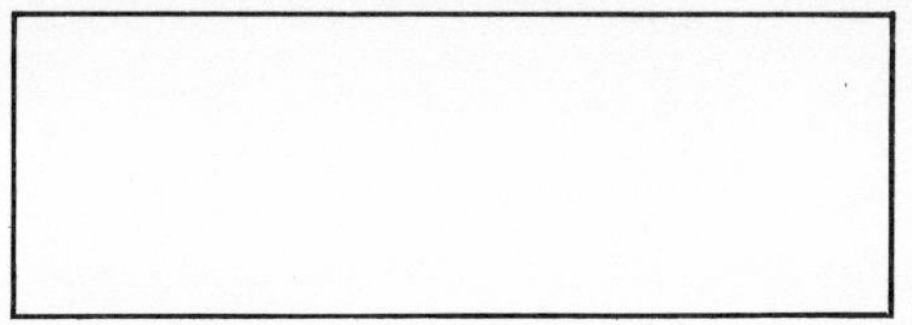

Length of lawn is 60 mm $\therefore$ true length is $\frac{60}{2}$ m = 30 m.

Width of lawn is 20 mm $\therefore$ true width is $\frac{20}{2}$ m = 10 m.

Example 2: Triangle ABC has to be enlarged so that the new figure has each side increased by a scale factor

a of 3 **b** of $\frac{1}{2}$.

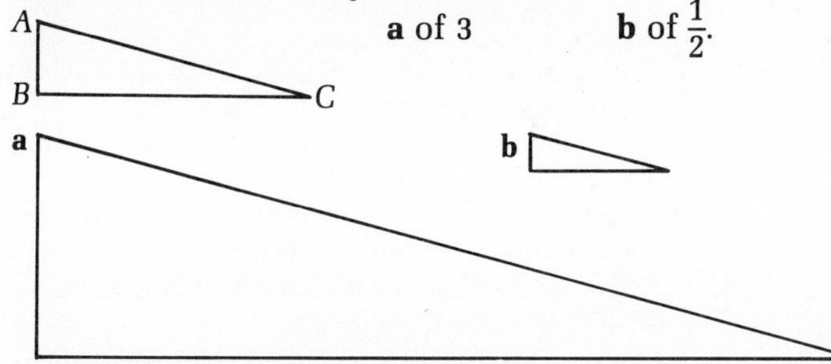

Example 3: Cambridge is on a bearing of 080° from Hereford and the distance between them is 225 km.

Guildford is on a bearing of 120° from Hereford and the distance between them is 195 km. Make a scale drawing of the information and find

a the distance from Guildford to Cambridge

b the bearing of Cambridge from Guildford

c the bearings of Hereford and Guildford from Cambridge.

A rough sketch might help.

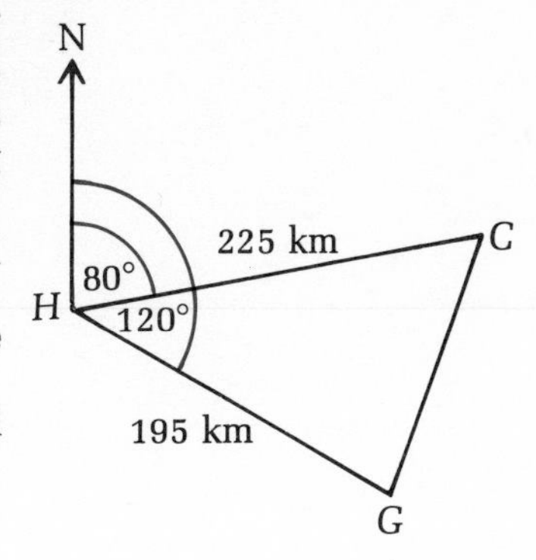

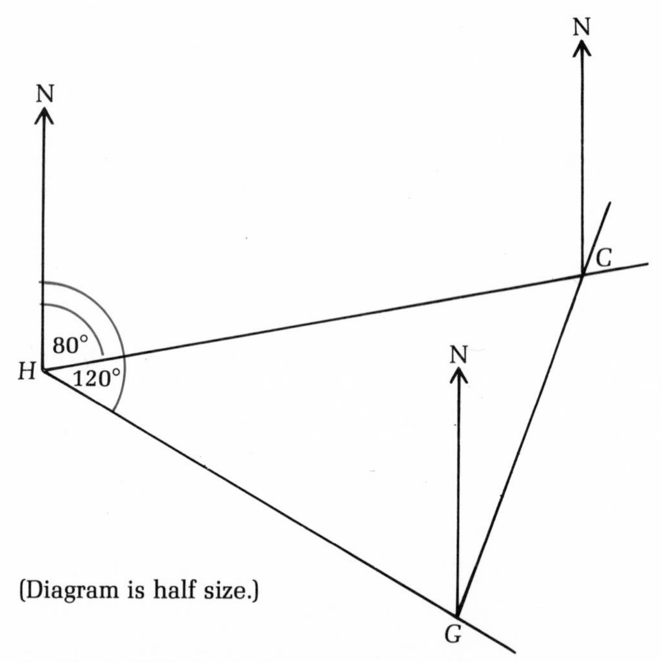

(Diagram is half size.)

Choose a suitable scale, say 1 cm to 15 km.
Select a point *H* for Hereford. Draw a North line at *H*. With your protractor measure the bearings of 80° and 120° and draw lines on which Cambridge and Guildford must lie. Measure 15 cm on the 080° bearing and 13 cm on the 120° bearing to fix *C* and *G*, respectively. Join *GC*.

a Measure *GC*: it is approximately 9.9 cm.
$\therefore$ distance from Guildford to Cambridge
$\simeq 9.9 \times 15 = 148.5$ km.

b Draw a North line at Guildford and measure bearings.
Bearing of Cambridge from Guildford is 020°.

c Bearing of Hereford from Cambridge is 260°.
Bearing of Guildford from Cambridge is 200°.

6 Symmetry and Tessellation

Foundation Level
Line symmetry, rotational symmetry, tessellations.

Intermediate Level
Sum of angles for tessellation, symmetry between two tangents of a circle from the same external point.

Foundation Level

At this level, you are concerned with two types of symmetry – line and rotational.

Line symmetry

If an isosceles triangle is folded along the dotted line *AB*, the two halves would fit over each other exactly.

The isosceles triangle is said to be symmetrical about the line of symmetry *AB*. Alternatively, *AB* is said to be the axis of symmetry.

A rectangle has two lines of symmetry.

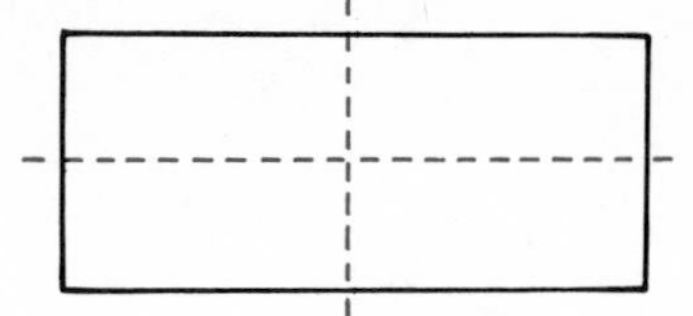

An equilateral triangle has three lines of symmetry.

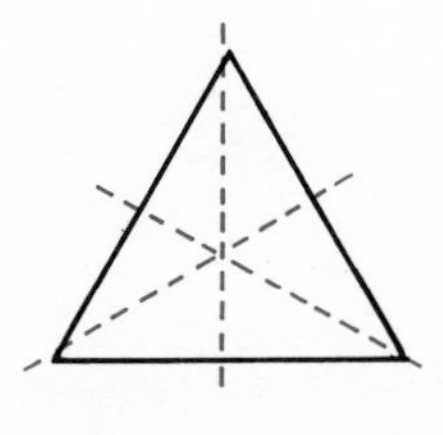

A square has four lines of symmetry.

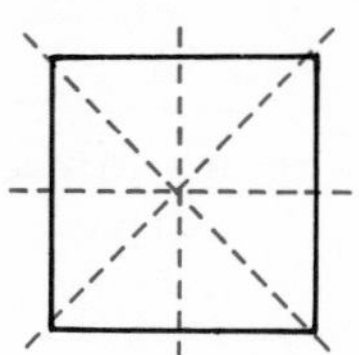

A circle has an infinite number of lines of symmetry, each being a diameter of the circle.

In the examination, you might be asked to draw the lines of symmetry of a figure or to find how many lines there are.

Rotational symmetry

This is rotating a figure about its centre so that it fits exactly the original position.

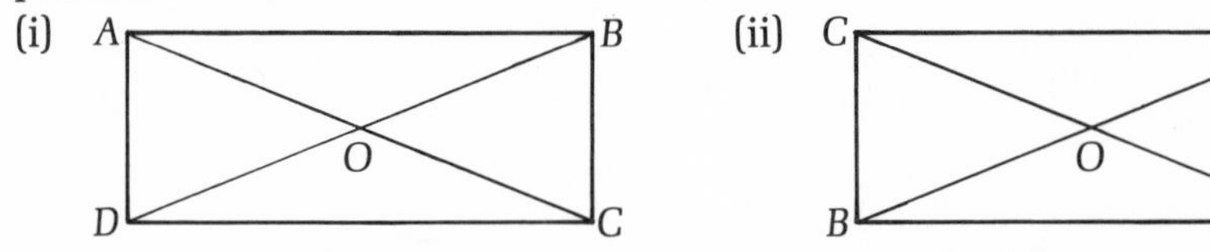

If the rectangle *ABCD* in figure (i) is rotated through 180° about *O*, it would fit exactly over the original *ABCD* but as shown in figure (ii). When *CDAB* is rotated through 180° again, the rectangle would be back in its original position. In this case, the rectangle is said to have rotational symmetry of order 2.

A square could be rotated 90° about its centre, a second 90° rotation about the centre, a third 90° rotation about the centre and finally a fourth 90° rotation about the centre before returning to its original position. A square has rotational symmetry of order 4.

Some other examples are:

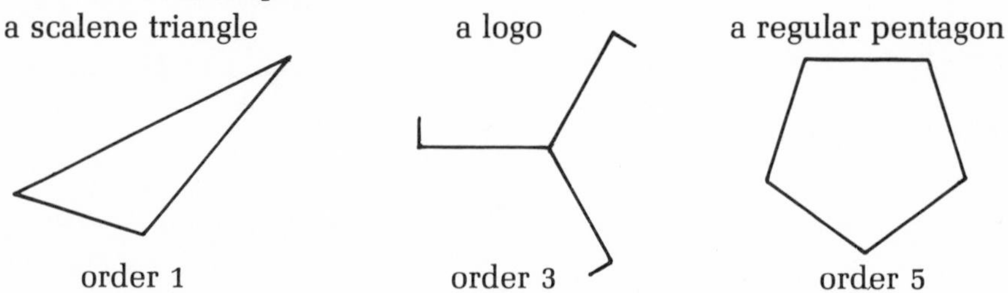

In the examination, you will be asked to recognise the order of rotational symmetry.

Tessellations

Shapes which fit together on a flat surface without leaving gaps are said to tessellate. At any point where tessellating shapes meet, the sum of the angles of the shapes at that point must be 360°.

Examples:

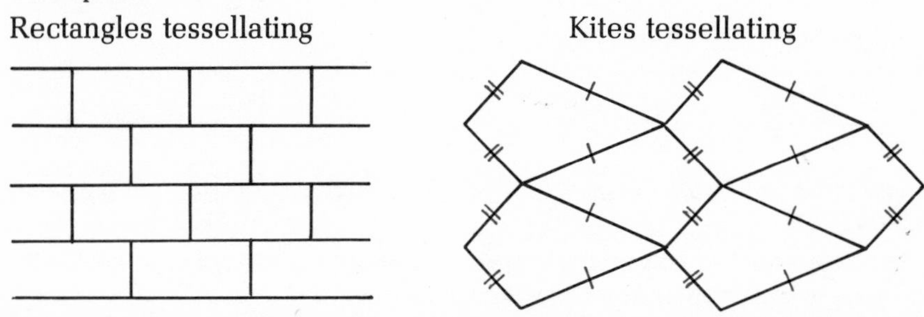

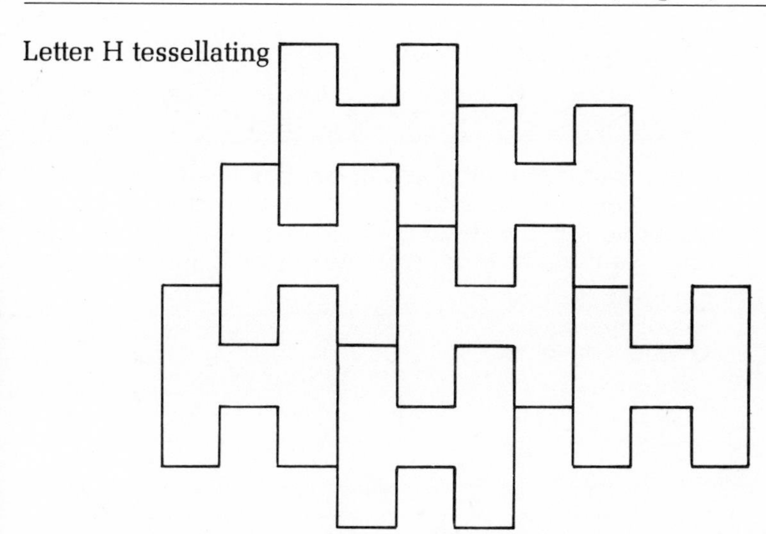

Some shapes made by two or more polygons form tessellations, e.g. a regular octagon and a square.

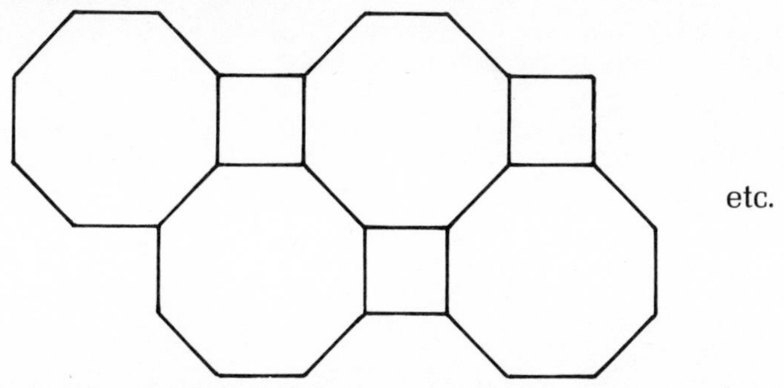

Intermediate Level

Sum of angles for tessellation

Not all regular polygons will tessellate. It is important to remember that if the shapes are to tessellate, the sum of the angles must be 360° at a point where shapes meet.

Example 1: Will regular hexagons tessellate?

Each interior angle of a regular hexagon is 120°. Since $\frac{360^\circ}{120^\circ}$ is 3, then regular hexagons will tessellate by 3 meeting at any junction.

This is a very common pattern.

Example 2: Will regular pentagons tessellate?

Each interior angle of a pentagon is 108°. $\frac{360°}{108°}$ does not give a whole number, so regular pentagons will not tessellate.

Some shapes have line symmetry only, or rotational symmetry only. There are other shapes which have both line and rotational symmetries. If we know something about line or rotational symmetries for a polygon, then we know some property of the polygon and vice versa.

A regular polygon of n sides will have n lines of symmetry and rotational symmetry of order n.

For instance, a regular octagon has 8 lines of symmetry as shown.

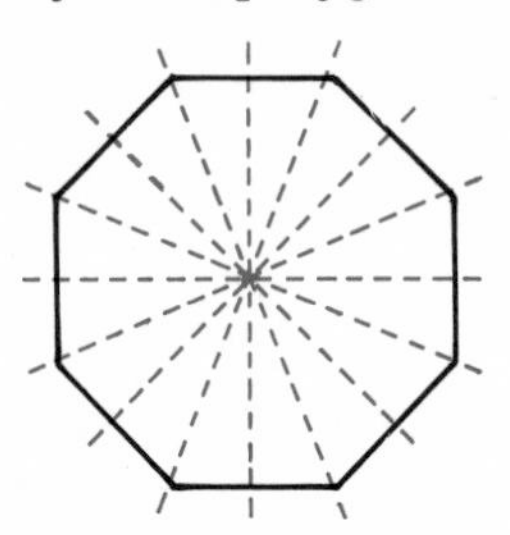

Tangents to a circle

If a tangent is drawn to a circle from an external point, there is always another tangent of equal length that can be drawn from the external point to the circle. The line joining the external point to the centre of the circle is always the line of symmetry.

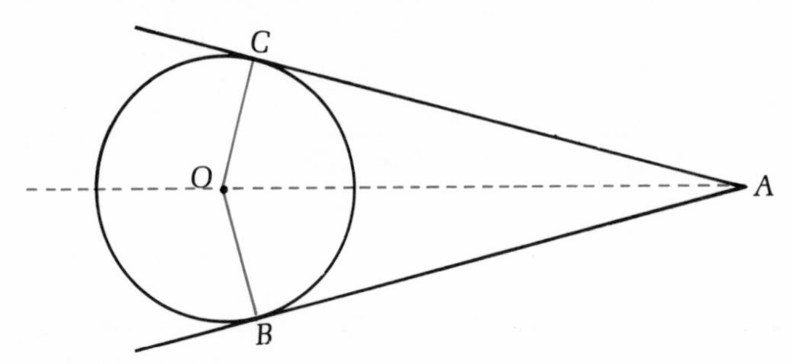

If O is the centre of the circle and A the external point, then OA is the line of symmetry.

The lengths of the tangents AB, AC are equal and $\angle\ OAC = \angle\ OAB$; also $\angle\ BOA = \angle\ COA$.

We already know that $\angle\ OBA = \angle\ OCA = 90°$.

7 Pythagoras's Theorem

Intermediate Level
The theorem.

Higher Level
Solution by algebra.

Intermediate Level

Pythagoras's theorem

The theorem states that in any right-angled triangle, the square of the length of the hypotenuse is equal to the sum of the squares of the lengths of the sides containing the right angle.

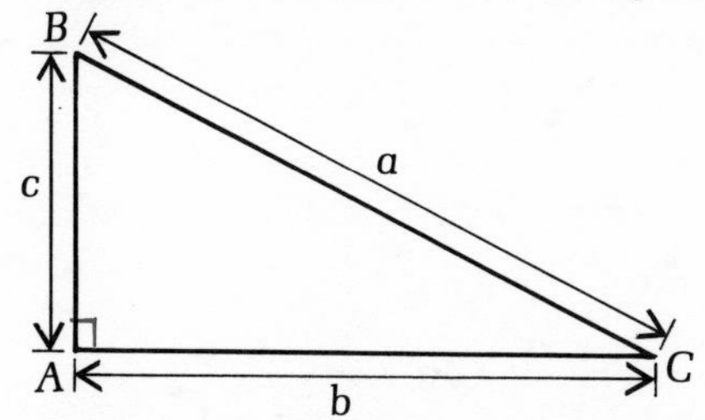

i.e. $BC^2 = AC^2 + AB^2$
or $a^2 = b^2 + c^2$

NOTE: In the first notation BC^2 means $BC \times BC$ not $B \times C^2$. It might be better to write $(BC)^2 = (AC)^2 + (AB)^2$.

In the examination, you will encounter numerical examples in which any two sides of a right-angled triangle are given and you are required to find the third side.

Example 1: In the diagram find a, correct to 2 decimal places.

$a^2 = 3^2 + 7^2 = 9 + 49 = 58$
$a = \sqrt{58} = 7.6157731$
$\therefore a = 7.62$ cm (to 2 d.p.)

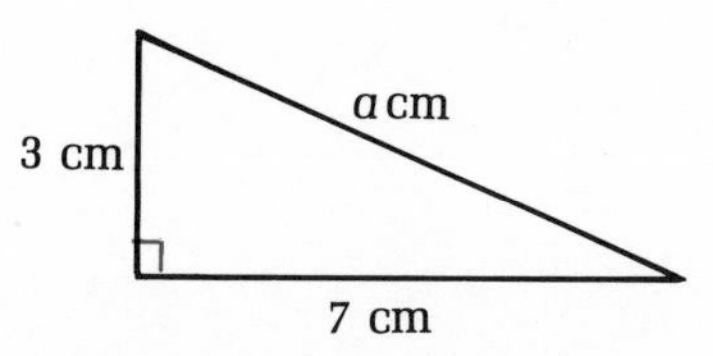

Example 2: In a triangle ABC, $AB = 8$ cm, $AC = 5$ cm and angle $C = 90°$.
Find the length of BC, correct to 3 s.f.
Sketch the diagram.

$(BC)^2 + 5^2 = 8^2$
$(BC)^2 + 25 = 64$
$(BC)^2 = 64 - 25 = 39$
$\therefore BC = \sqrt{39} = 6.244998$ cm $= 6.24$ cm (to 3 s.f.)

Higher Level

Solution by algebra

At this level, questions will be harder examples and will generally be solved more easily by the use of algebra.

Example: In the triangle ABC, $AB = 11$ cm, $AC = 5$ cm, and $BC = 8$ cm.

Find the lengths of BD, AD, and hence CD. Give the answers to 2 d.p.

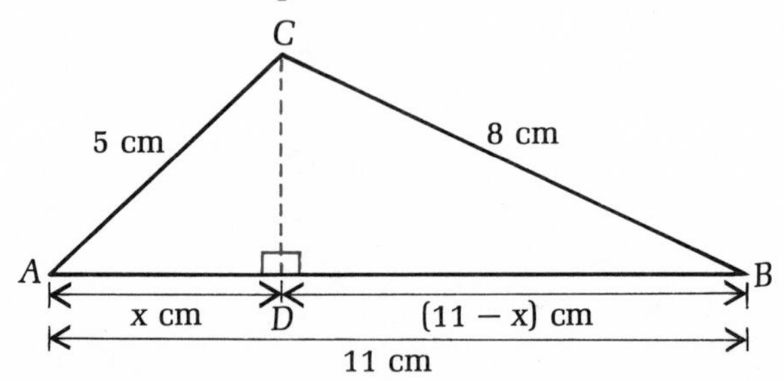

Let $AD = x$ cm, $\therefore DB = (11 - x)$ cm.

In $\triangle$ ADC, $(CD)^2 + x^2 = 5^2$, $\therefore (CD)^2 = 25 - x^2$

In $\triangle$ BDC, $(CD)^2 + (11 - x)^2 = 8^2$, $\therefore (CD)^2 = 64 - (11 - x)^2$

$$\therefore \; 25 - x^2 = 64 - (121 - 22x + x^2)$$

$$25 - x^2 = 64 - 121 + 22x - x^2$$

$$25 = -57 + 22x$$

$$25 + 57 = 22x$$

$$82 = 22x$$

$$\therefore x = \frac{82}{22} = 3.7272727$$

$\therefore AD = 3.73$ (to 2 d.p.)

$\therefore BD = 11 - 3.7272727 = 7.2727273 = 7.27$ cm (to 2 d.p.)

$\therefore (CD)^2 = 25 - (3.7272727)^2 = 11.107438$

$\therefore CD = 3.3327823$

$= 3.33$ cm (to 2 d.p.)

8 Trigonometry

Intermediate Level

Trigonometry of right-angled triangles, trigonometric ratios (tan, sin, cos), angle of elevation, angle of depression, use of bearings and trigonometry.

Higher Level

Sines and cosines of obtuse angles, graphs of sin x and cos x, sine rule, cosine rule, area of a triangle.

Intermediate Level

Trigonometry of a right-angled triangle

At this level, you are concerned only with the trigonometry of a right-angled triangle. Before defining any trigonometric ratio it is important to name the three sides of a right-angled triangle with reference to the acute angles of the triangle.

The side opposite the right angle is called the hypotenuse, *i.e.* AC.

The side opposite angle A is called the opposite side, *i.e.* BC.

The third side is called the adjacent side, *i.e.* AB adjacent to angle A.

If we consider angle C, AC is still the hypotenuse, but the side AB opposite angle C is now the opposite side and BC is the adjacent side.

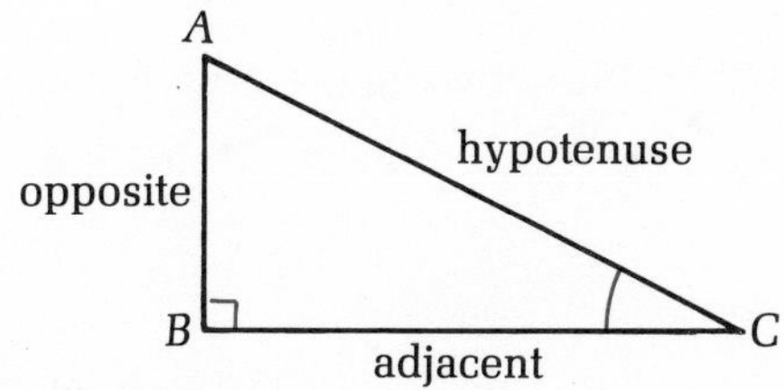

If you are given some information about one of the acute angles (or if you need to find an acute angle),

firstly mark the hypotenuse,
then the opposite side against the angle under consideration,
and finally the adjacent side.

Trigonometric ratios

You need to consider only three trig. ratios of acute angles in right-angled triangles.

Definitions

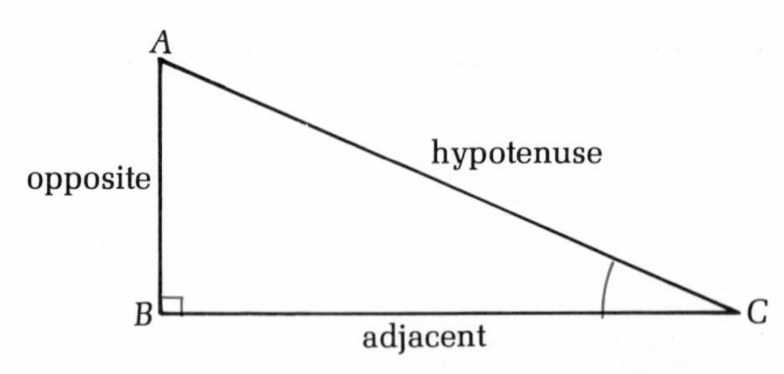

The *tangent* of angle C is the ratio $\dfrac{\text{length of } OPPOSITE \text{ side}}{\text{length of } ADJACENT \text{ side}}$.

Write, for short, $\tan C = \dfrac{AB}{BC}$.

The *sine* of angle C is the ratio $\dfrac{\text{length of } OPPOSITE \text{ side}}{\text{length of } HYPOTENUSE}$.

Write $\sin C = \dfrac{AB}{AC}$.

The *cosine* of angle C is the ratio $\dfrac{\text{length of } ADJACENT \text{ side}}{\text{length of } HYPOTENUSE}$.

Write $\cos C = \dfrac{BC}{AC}$.

The above three results are used to solve two different types of right-angled triangles.

Type A

When one acute angle and one side length are given, you can find the lengths of the other sides. The third angle is easily found by the angle sum (180°) of a triangle.

Type B

When two side lengths are given, you can find one of the acute angles. Hence the other acute angle can easily be found.

Example 1: In the diagram, find AB given that angle $C = 39°$, angle $B = 90°$, and $BC = 6$ cm.

Mark the hypotenuse, then the opposite side, and finally the adjacent side.

You know the adjacent side. You are required to find the opposite side, so use the definition of the tangent of angle C.

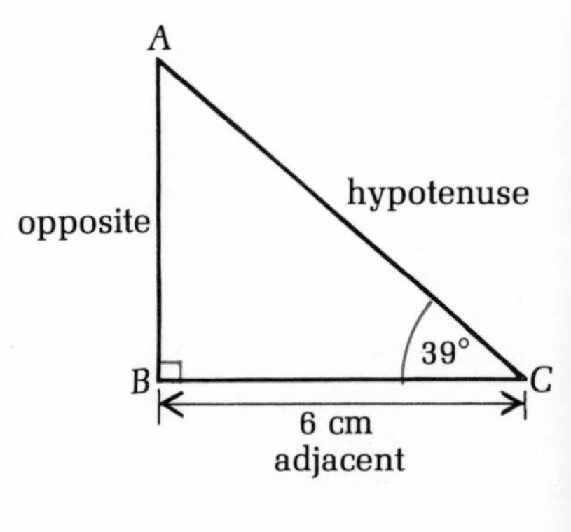

$$\tan C = \frac{AB}{BC}$$

$$\tan 39° = \frac{AB}{6}$$

$\therefore AB = 6 \times \tan 39° = 6 \times 0.809784\ldots$ cm (from calculator*)
$= 4.8587042$ cm

Normally you would be given some degree of accuracy in which to express the answer, e.g. 4.86 cm (to 3 s.f.).

Example 2: Suppose AC had been 6 cm instead of BC being known.

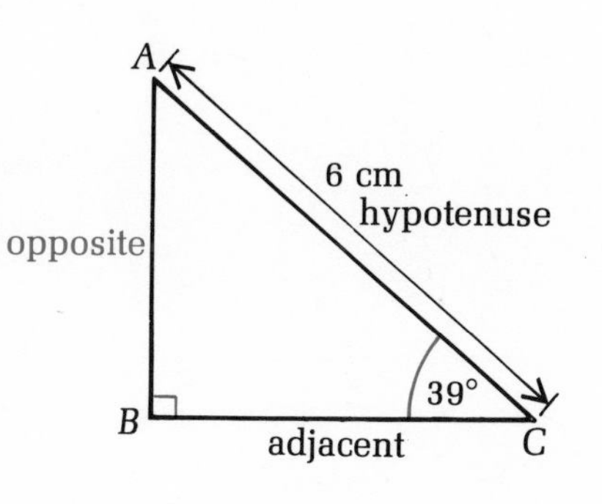

In this case you need to find the opposite side. You know the hypotenuse. Therefore, use the definition of the sine of $\angle C$.

$\sin C = \frac{AB}{6}$

$\sin 39° = \frac{AB}{6}$

$6 \times \sin 39° = AB$

$AB = 6 \times 0.6293203\ldots$ cm (from calculator*)

$= 3.7759223$ cm

$= 3.78$ cm (to 3 s.f.)

In the following examples, you are given the lengths of two sides of a right-angled triangle and are required to find a particular angle.

Example 3: In a triangle XYZ, $\angle X$ is 90°, $XZ = 15$ cm, $YX = 8$ cm. Find $\angle Y$.

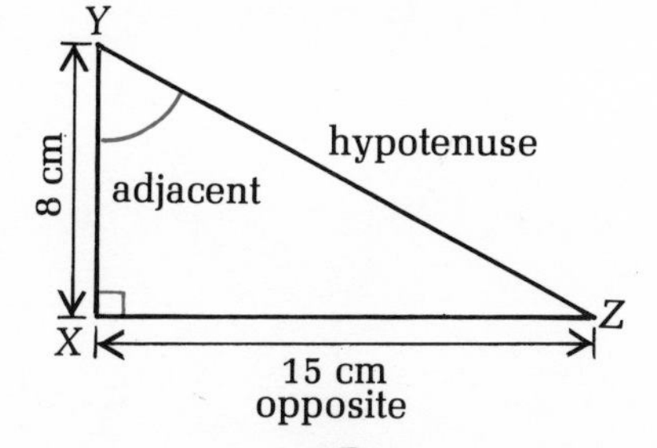

Clearly, $\tan Y = \frac{XZ}{YX} = \frac{15}{8}$
$= 1.875$

$\angle Y = 61.927513\ldots°$ (from calculator**)

$= 61.9°$ (to 3 s.f.)

Example 4: In a triangle PQR, $\angle P = 90°$, $PQ = 8.5$ cm, $QR = 12$ cm. Find $\angle Q$.

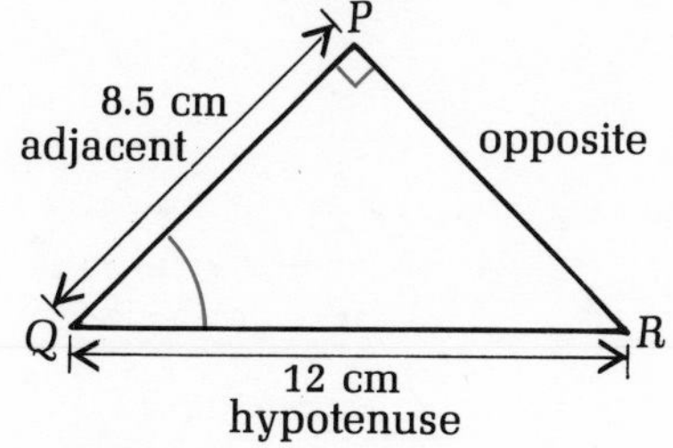

Clearly, $\cos Q = \frac{8.5}{12}$

$= 0.7083333\ldots$

$\therefore \angle Q = 44.900527\ldots°$ (from calculator**)

$= 44.9°$ (to 3 s.f.)

N.B. Check the calculations with your own calculator for these four examples.

* For these two, make sure that your calculator is in the degree mode, display the given angle (39), press appropriate trig. function button, then multiply by 6 to get answer.

** In the last two, make sure that your calculator is in the degree mode, do the appropriate division and use the inverse function key before pressing the appropriate trig. button.

Questions might be asked on problems involving angles of depression and elevation, and on the use of bearings.

Angle of elevation

When any object is observed by looking upwards, the angle formed between the object and the horizontal through your eye-level is called the angle of elevation.

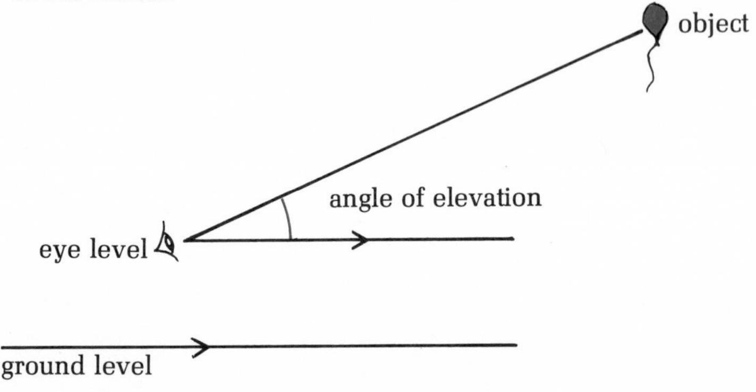

Angle of depression

This is similar to the angle of elevation, only this time an object is observed by looking downwards.

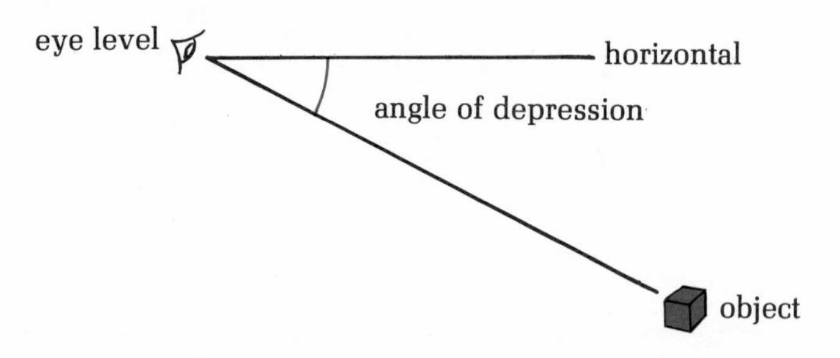

Example 1: The angle of elevation of a flag-pole is 32° from a point on level ground, 18 m away from the base of the flag-pole. Find the height of the flag-pole.

Let AB be the flagpole, and C be the point 18 m away from B.

$$\frac{AB}{18} = \tan 32^\circ$$

$$\begin{aligned} AB &= 18 \times \tan 32^\circ \\ &= 18 \times 0.6248693 \text{ m} \\ &= 11.247648 \text{ m} \\ &= 11.2 \text{ m (3 s.f.)} \end{aligned}$$

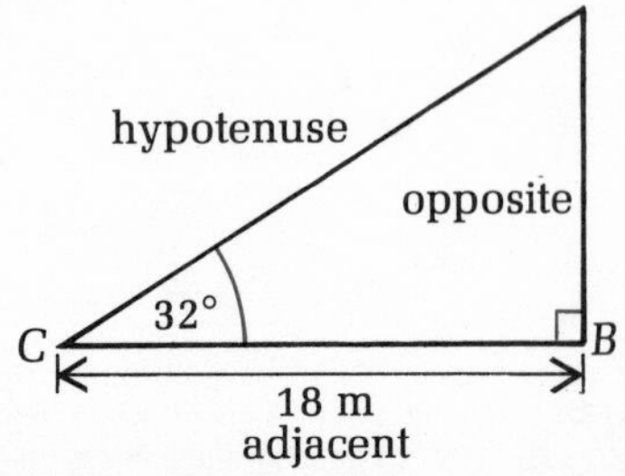

Example 2: A boat is observed from the top of a cliff 100 m above sea-level and the angle of depression is known to be 41°. How far is the boat from the foot of the vertical cliff?

Let AB be the cliff and C the boat.

By alternate angles, $\angle C = 41°$

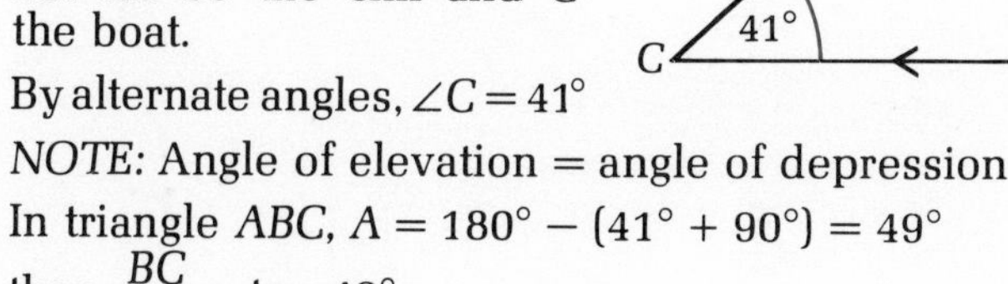

NOTE: Angle of elevation = angle of depression

In triangle ABC, $A = 180° - (41° + 90°) = 49°$

then $\frac{BC}{100} = \tan 49°$

$BC = 100 \times \tan 49° = 115.03684\ldots$ m

$= 115$ m (to 3 s.f.)

Use of bearings and trigonometry

Example: A ship sails from a port P to an island Q, 70 km away on a bearing of 44°. How far North has the ship sailed?

Draw a North line from P.

Let PQ represent the path of the ship.

Draw QR at right angles to the North line through P.

Then PR is the required distance.

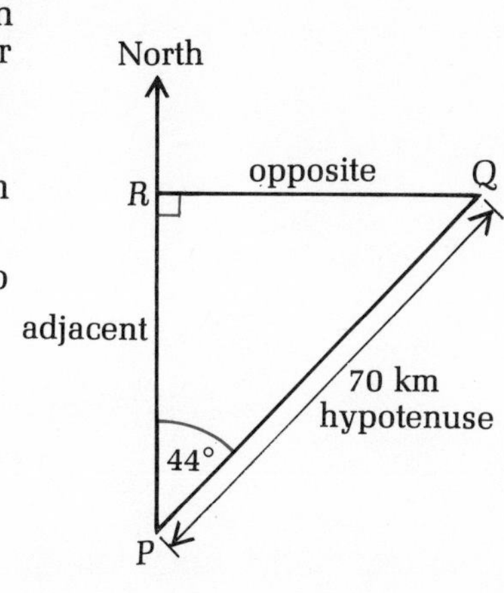

$\frac{PR}{70} = \cos 44°$

$PR = 70 \times \cos 44°$ km

$= 70 \times 0.7193398$

$= 50.353786$

$= 50.4$ km (to 3 s.f.)

If you were asked to find the distance travelled East, you need to find QR.

$\frac{QR}{70} = \sin 44°$

$QR = 70 \times \sin 44°$ km

$= 70 \times 0.6946583$

$= 48.626086$

$= 48.6$ km (to 3 s.f.)

Higher Level

Sines and cosines of obtuse angles

The definition of trigonometric ratios must now be extended to find sines and cosines of obtuse angles. It is obvious that there cannot be an obtuse angle in a right-angled triangle, so you must redefine the initial definitions of sine and cosine.

Draw a semicircle with centre O on rectangular axes. Let OP be a radius free to rotate from OX to OX'. Draw perpendiculars from P to OX and OY at M and N, respectively. Let $\angle POM = \theta$ at any instant, measuring θ anticlockwise from OA.

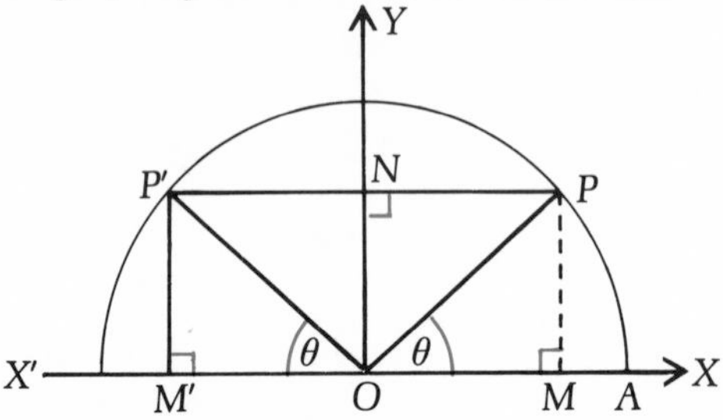

$$\text{Then } \sin\theta = \frac{PM}{OP} \qquad \cos\theta = \frac{OM}{OP}$$

By symmetry, $PM = ON$

$$\sin\theta = \frac{ON}{OP} = \frac{\text{projection of } OP \text{ on to } y\text{-axis}}{OP}$$

$$\cos\theta = \frac{\text{projection of } OP \text{ on to } x\text{-axis}}{OP}$$

Now use these definitions in the next quadrant for obtuse angles. Draw PNP' parallel to the x-axis, $P'M'$ is perpendicular to x-axis. By symmetry, $P'OM' = \theta$, $\therefore \angle P'OA$ is $180° - \theta$

$$\sin P'OA = \frac{\text{projection of } OP' \text{ on to } y\text{-axis}}{OP} = \frac{ON}{OP}$$

i.e. $\sin(180° - \theta) = \sin\theta$

$$\cos P'OA = \frac{\text{projection of } OP' \text{ on to } x\text{-axis}}{OP} = \frac{OM'}{OP}$$

But OM' is the same length as OM by symmetry but in the negative direction.

$$\cos(180 - \theta) = \frac{-OM}{OP} = -\cos\theta$$

or multiplying by -1, $\cos\theta = -\cos(180 - \theta)$

SUMMARY Sin $(180° - \theta) = \sin\theta$ cos $(180° - \theta) = -\cos\theta$

Graphs of sin x and cos x

NOTE: In the range $0° - 180°$, $\sin x°$ is always positive. Its least value is 0 at 0° and 180°, its greatest value is 1 when x is 90°.

Cos x is positive for acute angles going from 1 to 0, negative for obtuse angles going from 0 to -1 as x goes from 90° to 180°.

The graph of $y = \sin x$ has reflective symmetry about the line $x = 90°$.

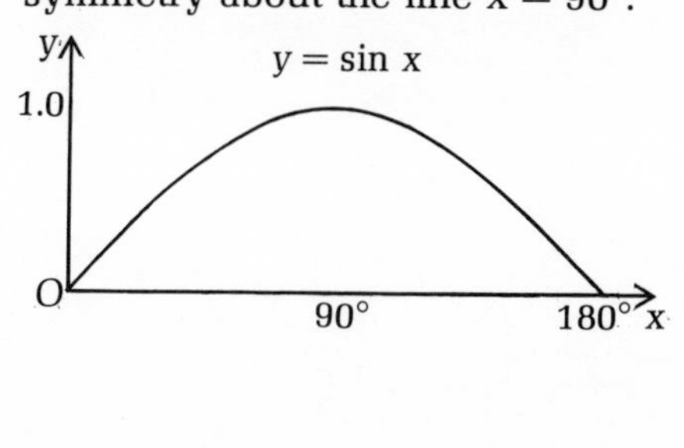

The graph of $y = \cos x$ has rotational symmetry about $x = 90°$.

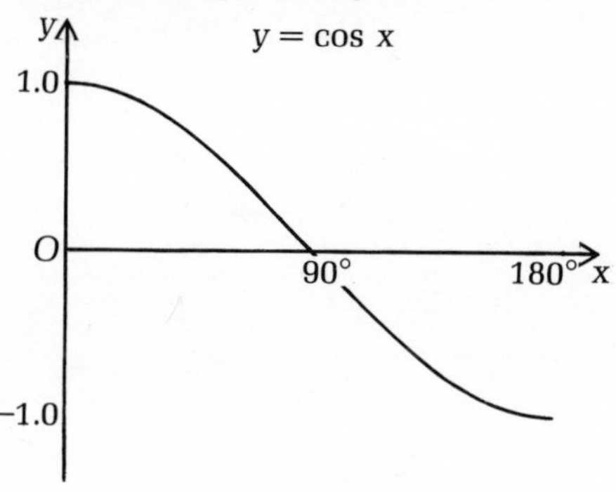

The graphs can help you in solving problems and their general shape should be remembered.

Example: $\sin 120° = \sin (180° - 120°) = \sin 60°$

or $\sin 145° = \sin (180° - 145°) = \sin 35°$

These results can also be verified by using your calculator.

$\cos 120° = -\cos (180° - 120°) = -\cos 60°$

$\cos 145° = -\cos (180° - 145°) = -\cos 35°$

Sine rule

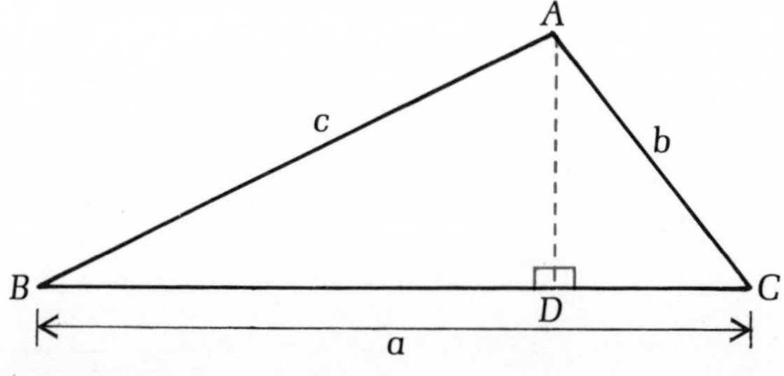

In any triangle, a common notation is to letter the vertices with capital letters (e.g. *A*, *B*, and *C* above) and to denote the sides opposite the vertices with corresponding small letters. Thus $a = BC$, $b = AC$, $c = AB$.

Construct a perpendicular from *A* to *BC* at *D*.

In $\triangle ABD$, $\dfrac{AD}{c} = \sin B$ $\quad \therefore AD = c \sin B$ ①

In $\triangle ACD$, $\dfrac{AD}{b} = \sin C$ $\quad \therefore AD = b \sin C$ ②

From ① and ②, $b \sin C = c \sin B$

Rearranging: $\dfrac{b}{\sin B} = \dfrac{c}{\sin C}$.

Similarly if a perpendicular is drawn from C to AD it can be shown that

$$\frac{a}{\sin A} = \frac{b}{\sin B} \quad \text{and} \quad \frac{a}{\sin A} = \frac{b}{\sin B} = \frac{c}{\sin C}$$

If the circumcircle is drawn through A, B and C it can be shown that

$$\frac{a}{\sin A} = \frac{b}{\sin B} = \frac{c}{\sin C} = 2R$$

where R is the radius of the circumcircle.

SUMMARY $\quad \dfrac{a}{\sin A} = \dfrac{b}{\sin B} = \dfrac{c}{\sin C} = 2R$

The sine rule is used in solving triangles that do not have a right angle when
2 angles and any side are known,
or 2 sides and an angle opposite one of the sides are known.

Example 1: In the diagram find angle B.
In this case write the sine rule

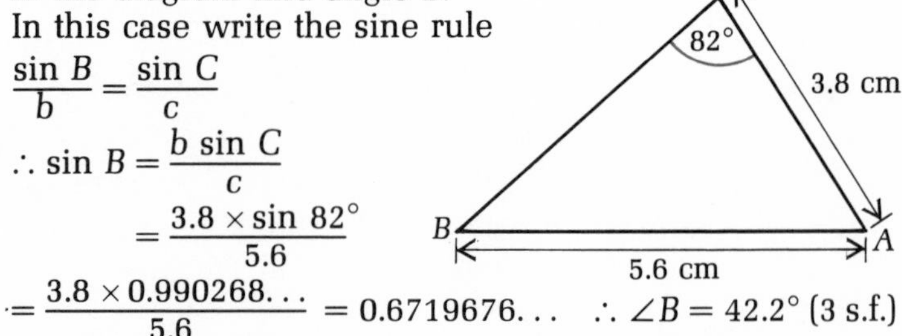

$$\frac{\sin B}{b} = \frac{\sin C}{c}$$

$$\therefore \sin B = \frac{b \sin C}{c}$$

$$= \frac{3.8 \times \sin 82^\circ}{5.6}$$

$$= \frac{3.8 \times 0.990268\ldots}{5.6} = 0.6719676\ldots \quad \therefore \angle B = 42.2^\circ \text{ (3 s.f.)}$$

You could now find angle A and hence a by the same rule. The triangle could thus be completely solved.

Example 2: Find AC using this diagram.

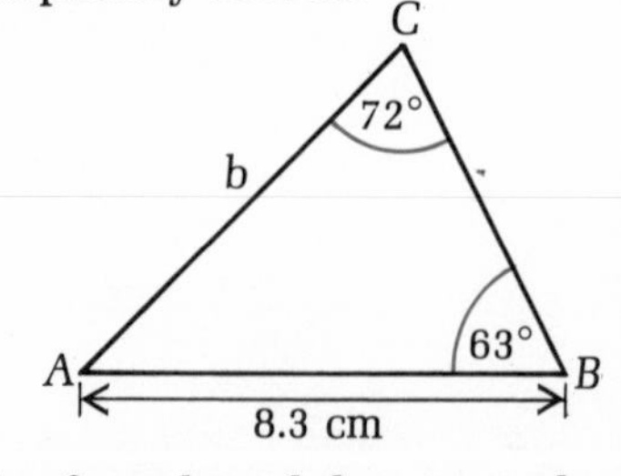

$$\frac{b}{\sin B} = \frac{C}{\sin C}$$

$$\therefore b = \frac{8.3 \sin 63^\circ}{\sin 72^\circ}$$

$$= 7.7759355\ldots$$

$$= 7.78 \text{ cm (to 3 s.f.)}$$

Clearly, angle A could easily be found, and the sine rule applied again if you wished to find BC.

Cosine rule

In $\triangle ABC$ let CD be the perpendicular from C to AB at D.

Let $CD = h$, $AD = x$. $\quad \therefore BD = c - x$.

In $\triangle ACD$ by Pythagoras

$x^2 + h^2 = 6^2 \qquad \therefore h^2 = b^2 - x^2 \qquad$ ①

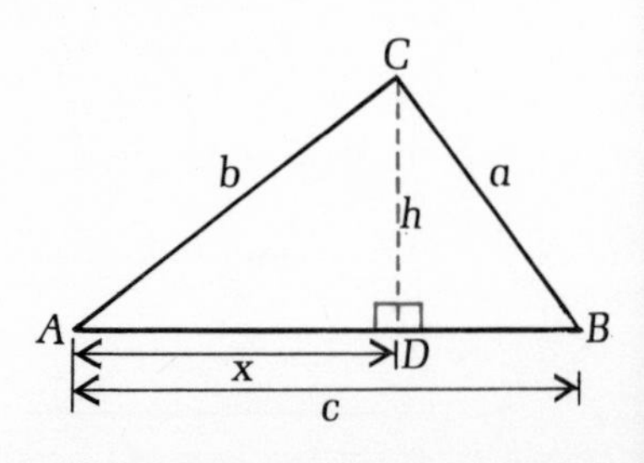

In $\triangle CBD$ similarly $\quad h^2 + (c - x)^2 = a^2 \quad h^2 = a^2 - (c - x)^2$ ②

From ① and ② $\quad \therefore a^2 - (c^2 - 2cx + x^2) = b^2 - x^2$

$$a^2 - c^2 + 2cx - x^2 = b^2 - x^2$$

$$\therefore a^2 = b^2 + c^2 - 2cx \quad ③$$

In $\triangle ACD$ $\quad \cos A = \frac{x}{b} \quad \therefore b \cos A = x$ ④

Substitute ④ in ③ $\quad a^2 = b^2 + c^2 - 2bc \cos A$

Likewise if perpendiculars are drawn from A to BC and B to AC, it can be shown that $b^2 = a^2 + c^2 - 2ac \cos B$
and $c^2 = a^2 + b^2 - 2ab \cos C$

NOTE: These are three distinct formulae and not transpositions of each other.

SUMMARY $\quad a^2 = b^2 + c^2 - 2bc \cos A$

The cosine rule is used in solving triangles that do not have a right angle when
3 sides are known
or 2 sides and the angle between them are known.

Example 1: In $\triangle ABC$, given $AB = 6$ cm, $BC = 3$ cm, $AC = 7$ cm, find $\angle B$.

If we need $\angle B$, we use $b^2 = a^2 + c^2 - 2ac \cos B$.

i.e. $\quad 7^2 = 3^2 + 6^2 - 2 \times 3 \times 6 \times \cos B$

$$49 = 9 + 36 - 36 \cos B$$

$$\therefore 36 \cos B = 9 + 36 - 49 = -4$$

$$\cos B = \frac{-4}{36}$$

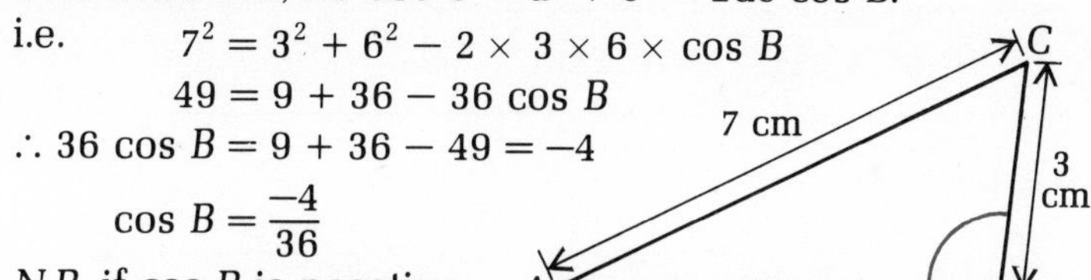

N.B. if $\cos B$ is negative, $\angle B$ must be obtuse.

$\angle B = 96.37937\ldots^\circ$ (from calculator)
$= 96.4^\circ$ (3 s.f.)

If you wish to solve the triangle completely, you could again use the cosine rule for either $\angle A$ or $\angle C$, but it would probably be easier this time to use the sine rule.

Example 2: If in $\triangle ABC$, $AB = 8$ cm, $AC = 6$ cm and $\angle A = 81^\circ$, find BC.

By cosine rule

$$a^2 = b^2 + c^2 - 2bc \cos A$$
$$= 6^2 + 8^2 - 2 \times 8 \times 6 \times \cos 81^\circ$$
$$= 36 + 64 - 15.017708\ldots$$
$$= 84.982291\ldots$$

$\therefore a = \sqrt{84.982291} \ldots$ cm $= 9.218584$ cm $= 9.22$ cm (to 3 s.f.)

If you needed to solve the triangle completely, you could again use the cosine rule knowing three sides to find $\angle B$ or $\angle C$, but it would be easier to use the sine rule to get $\angle B$ or $\angle C$. If you use the sine rule, it is better to find the smaller of the two angles, *i.e.* $\angle B$ (the angle opposite the smaller side).

Area of a triangle

The area of the triangle has already been stated to be

$$\frac{1}{2} \times \text{base} \times \text{perpendicular height.}$$

Consider the triangle ABC, where AD is the perpendicular height from A to D on BC.

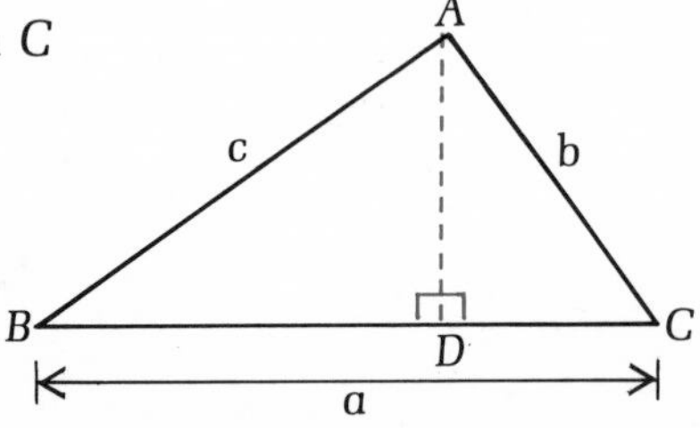

In $\triangle ACD$, $\sin C = \frac{AD}{b}$ $\quad \therefore AD = b \sin C$

$\therefore$ area of $\triangle ABC = \frac{1}{2} a \times b \sin C$

$= \frac{1}{2} ab \sin C$

Alternatively it can be shown that the area is $\frac{1}{2} ac \sin B$ or $\frac{1}{2} bc \sin A$.

Put into words, the area of any triangle is half the product of two sides multiplied by the sine of the included angle.

SUMMARY Area of $\triangle ABC = \frac{1}{2} ab \sin C$

Example:

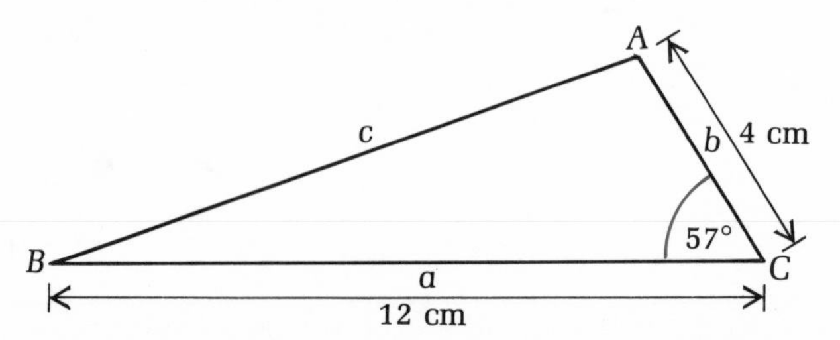

Area of $\triangle ABC = \frac{1}{2} ab \sin C = \frac{1}{2} \times 12 \times 4 \times \sin 57° \text{ cm}^2$

$= 20.128093\ldots \text{cm}^2$ (from calculator)

$= 20.1 \text{cm}^2$ (3 s.f.)

In the examination, questions could be set which involve simple applications to problems in three dimensions. It is not intended to show any here. Such questions need good clear diagrams and then you need to work carefully with the appropriate triangles.

9 Transformations

Intermediate Level
Reflections, rotations, translations.

Higher Level
Reflections, rotations, translations, enlargements, combination of transformations, combination of two reflections, two rotations, a rotation and a reflection.

Intermediate Level

In geometry, a transformation is the change of one figure to another figure. At this level, we are concerned with drawing images or naming the type of transformation which maps an original figure onto its image. There are three types of transformation to be considered, two of which were mentioned briefly in module 16.

Reflections

We are concerned only with reflections in horizontal and vertical lines. In the diagram, the triangle ABC is to be reflected in both the x-axis and the y-axis.

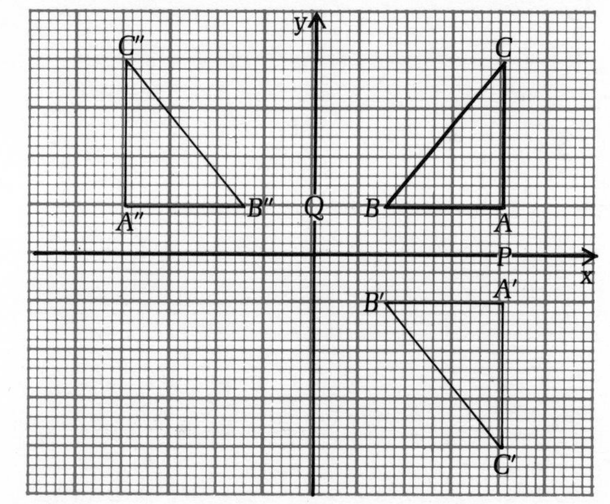

The triangle $A'B'C'$ is the image of triangle ABC when reflected in the x-axis.

The reflected image of triangle ABC in the y-axis is the triangle $A''B''C''$.

NOTE: In any reflection the size and shape of the image is unchanged. The line of reflection is the perpendicular bisector of the line joining any point on the given figure to its image point.

Thus the distances AP and $A'P$ are equal, CP and $C'P$ are equal, and BQ and $B''Q$ are equal.

In the next diagram, name the type of transformation which maps *ABCD* to *A′B′C′D′*.

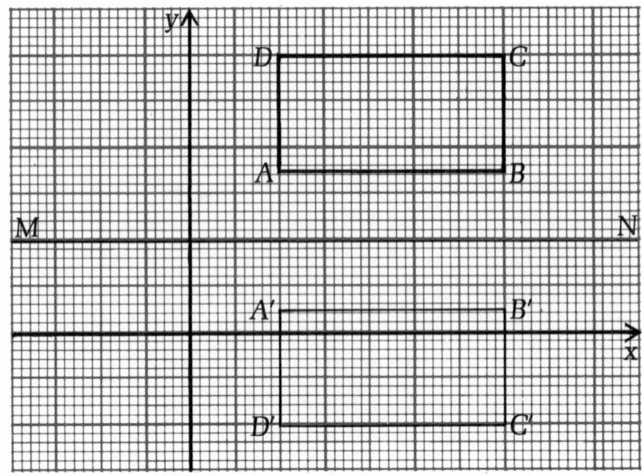

The transformation is a reflection in the line *MN*. If we have to name a reflection always give the line of reflection, either as above *MN* or its actual equation.

Rotations

The rotations to be considered will be those about the origin, vertices or midpoints of edges through angles which are multiples of 90°.

In the diagram, triangle *ABC* has been rotated clockwise about *O* through 90°, 180° and 270° to give images *A′B′C′*, *A″B″C″* and *A‴B‴C‴* respectively.

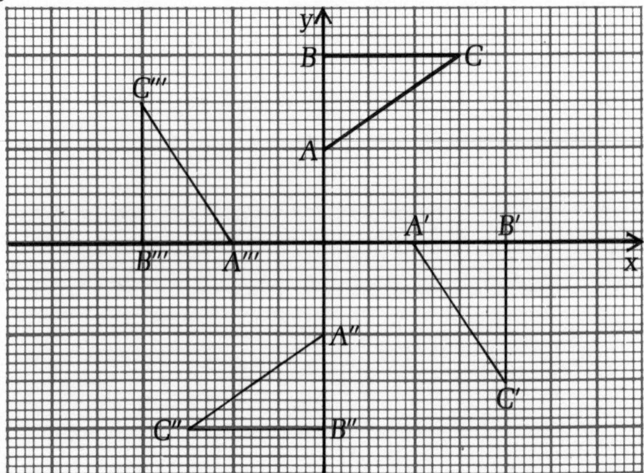

NOTE: Just as for reflections, the size and shape of the image is unchanged. The angle of rotation is the angle between the line joining any point on the given figure to the centre of rotation and the line joining the corresponding image point to the centre of rotation, e.g. in the diagram $\angle AOA' = \angle BOB' = \angle COC' = 90°$.

The rotation can either be clockwise or counter-clockwise (anticlockwise). A clockwise rotation is said to be negative and a counter-clockwise rotation is said to be positive.

Thus $A'B'C'$ is the image ABC when ABC is rotated about O through an angle of $-90°$, or when ABC is rotated about O through an angle of 270°.

In naming the type of transformation, we must state rotation, the centre of rotation and the angle (the direction or whether it is positive or negative).

Example: Name the type of transformation in the following diagram for the figure $ABCD$.

$AB'C'D'$ is the rotation of $ABCD$ about the centre of rotation A through a clockwise angle of 180°.

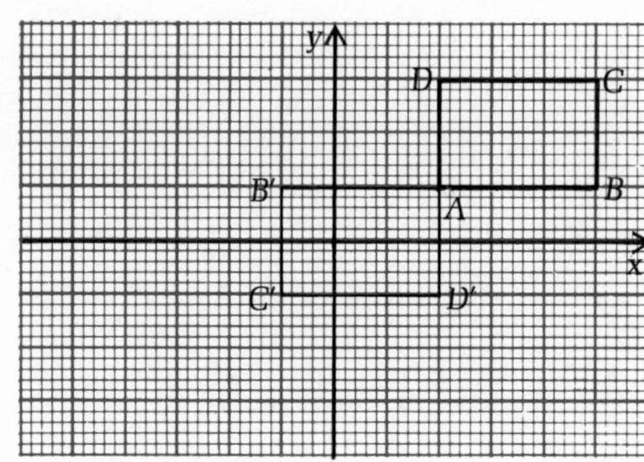

Translations

When every point in a plane figure moves the same distance in the same direction, the transformation is called a translation. A translation is sometimes called a *shift*.

In the figure below, $A'B'C'D'$ is a translation of $ABCD$. Likewise $A''B''C''D''$ is another translation of $ABCD$.

Again in a translation, the size and shape of the image are unchanged. Note that AA' is parallel and equal to BB', CC', DD', or AA'' is parallel and equal to BB'', CC'', DD''.

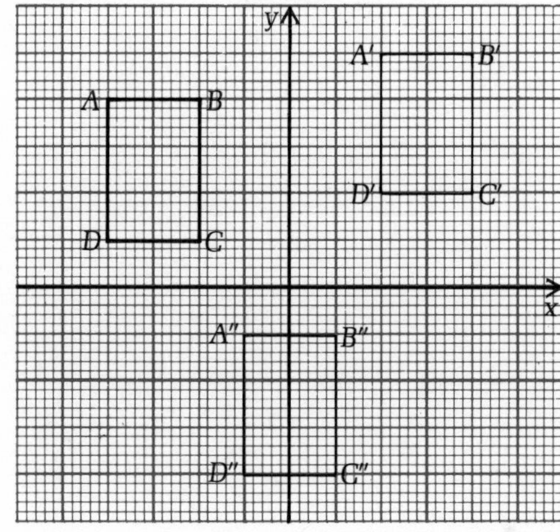

Higher Level

At this level you are concerned with the use of coordinates for sets of points and their images, together with further work on reflections, rotations and translations.

Reflections

Questions will normally involve reflections in the lines $y = x$ and $y = -x$, or in lines parallel to axes.

Examples: In the diagram, *A* is the point (3, 2). Find the images of *A* in the lines.

1 $x = 5$, **2** $y = -1$, **3** $y = x$, **4** $y = -x$.

1 Draw the line $x = 5$, then image *B* is (7, 2).
NOTE: For reflection in a line parallel to *y*-axis, there is no change in *y*-coordinate for the image.

2 Draw the line $y = -1$, then image *C* is (3, −4).
NOTE: For reflection in a line parallel to *x*-axis, there is no change in *x*-coordinate for the image.

3 Draw the line $y = x$. The image *D* is (2, 3).
NOTE: The line $y = x$ is the perpendicular bisector of *AD* and the effect of such a transformation is to reverse the coordinates of *A*.

4 Draw the line $y = -x$. The image *E* is (−2, −3).
NOTE: The line $y = -x$ is the perpendicular bisector of *AE* and the effect of such a transformation is to reverse both the coordinates of *A* and the signs of these coordinates.

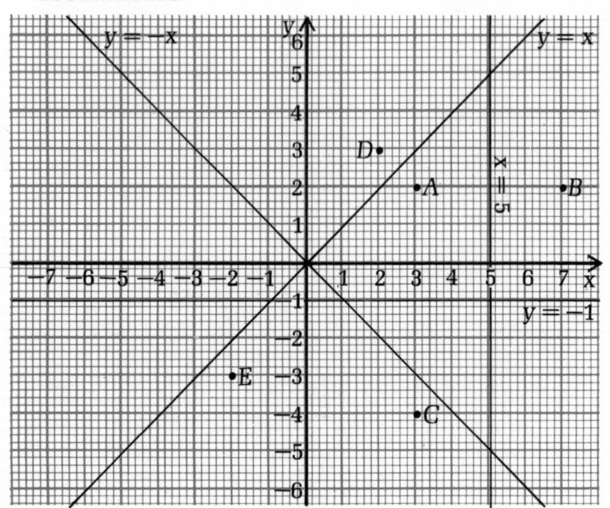

Rotations

At this level you are concerned with rotations about any point through some angle which will be a multiple of 90°.

Example: In the diagram, *P* is the point (1, 3) and *A* is the point (5, 1). Find the images of point *A* for rotations of 90°, 180° and 270° about the centre of rotation *P*.

One method is to draw a circle centre P, radius PA. Draw the diameter APC. Draw the perpendicular bisector of AC through P to meet the circle at B and D.

B is the image of A after a rotation of 90° about P, i.e. (3, 7)

C is the image of A after a rotation of 180° about P, i.e. (−3, 5)

D is the image of A after a rotation of 270° about P, i.e. (−1, −1)

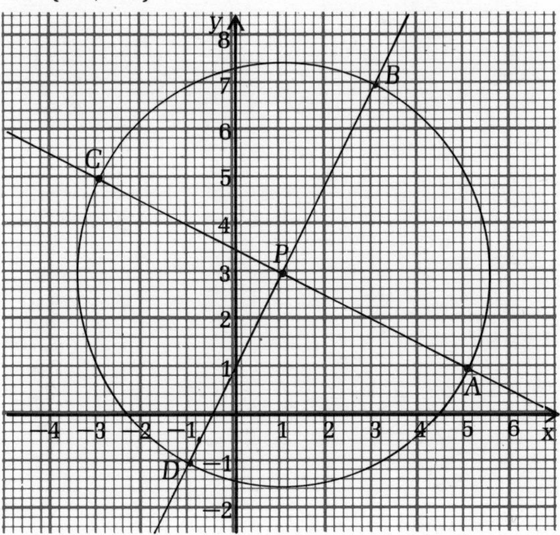

Alternatively, the images can be found by similar triangles as shown in the following diagrams.

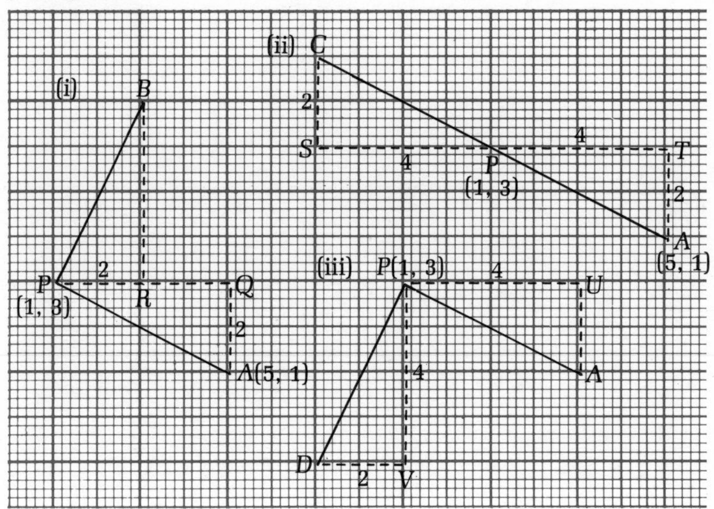

in fig. (i) $AQ = 2$, $PQ = 4$ $\therefore PR = 2$, $BR = 4$
$\therefore$ x-coordinate of B is $1 + 2 = 3$
y-coordinate of B is $3 + 4 = 7$ $\therefore B$ is $(3, 7)$

in fig. (ii) $AT = 2$, $PT = 4$ $\therefore SC = 2$, $SP = 4$
$\therefore$ x-coordinate of C is $1 - 4 = -3$
y-coordinate of C is $3 + 2 = 5$ $\therefore C$ is $(-3, 5)$

in fig. (iii) $AU = 2$, $PU = 4$ $\therefore DV = 2$, $VP = 4$
$\therefore$ x-coordinate of D is $1 - 2 = -1$
y-coordinate of D is $3 - 4 = -1$ $\therefore D$ is $(-1, -1)$

Translations

The shift, described earlier, is usually expressed in the form of a column vector (or displacement vector).

In the diagram below, the point A is transformed to point B by moving from (1, 2) to (6, 5), *i.e.* the shift is 5 units to the right and 3 units upwards.

This shift is written as $\begin{pmatrix} 5 \\ 3 \end{pmatrix}$, *i.e.* in the form $\begin{pmatrix} x \\ y \end{pmatrix}$

where x is the shift parallel to x-axis, and
y is the shift parallel to y-axis.

Point A is given a translation of $\begin{pmatrix} 2 \\ -3 \end{pmatrix}$. Starting at A move 2 units to the right and 3 units downwards, *i.e.* at C.

Point A is given a translation of $\begin{pmatrix} -3 \\ -4 \end{pmatrix}$, *i.e.* image is $D \begin{pmatrix} -2 \\ -2 \end{pmatrix}$.

Point A is given a translation of $\begin{pmatrix} -5 \\ 0 \end{pmatrix}$, *i.e.* image is $E \begin{pmatrix} -4 \\ 2 \end{pmatrix}$.

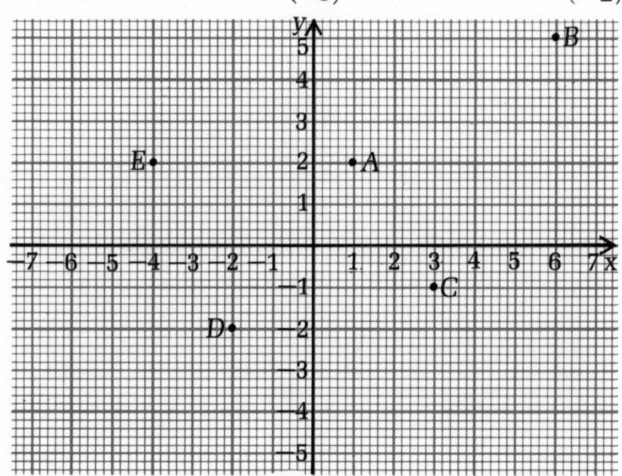

Enlargements

When all the lengths of a plane figure are multiplied by a scale factor (greater than 1), the transformation that is obtained is called an enlargement. If the scale factor is less than 1, the transformation is called a *reduction*.

An enlargement or reduction will involve a defined centre of enlargement, and for the purposes of the examination this will always be the origin.

Example 1: In the diagram, the square *OABC* is as shown. *OA′B′C′* is an enlargement with a scale factor of 2. *OA″B″C″* is a reduction with a scale factor of $\frac{1}{2}$.

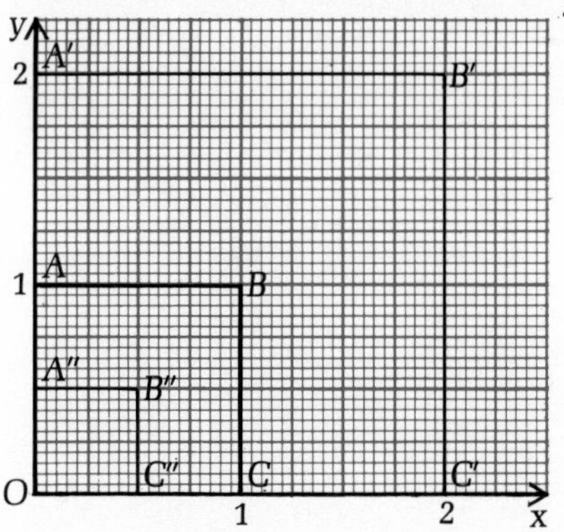

Example 2: To construct an enlargement of scale factor 2 and centre *O* of the triangle *ABC*, shown in the diagram,

join *OC* and produce to *C′* so that $OC' = 2 \times OC$
join *OA* and produce to *A′* so that $OA' = 2 \times OA$
join *OB* and produce to *B′* so that $OB' = 2 \times OB$

Triangle *A′B′C′* is the enlargement of triangle *ABC*, centre *O*.

To produce an enlargement of scale factor *n*, with origin as centre, simply take OA', $= n \times OA$, $OB' = n \times OB$, $OC' = n \times OC$.

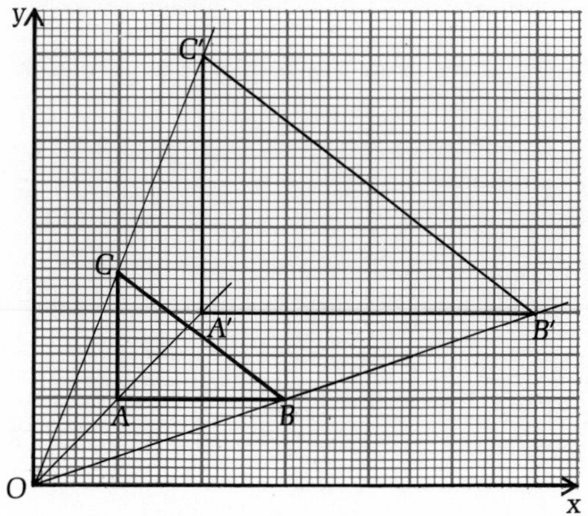

NOTE: With enlargements (and reductions) the images are similar figures to the given figures, corresponding angles being equal.
The area of the image is the area of the given figure multiplied by the (scale factor)^2.

In the last diagram $AB = 4$ units, $AC = 3$ units.

$\therefore$ area of $\triangle\ ABC \;= \frac{1}{2} \times 4 \times 3 = 6$ sq. units

$A'B' = 8$ units, $B'C' = 6$ units

$\therefore$ area of $\triangle\ A'B'C' = \frac{1}{2} \times 8 \times 6 = 24$ sq. units

Check: Area of $\triangle\ A'B'C' = 6 \times (2)^2 = 6 \times 4 = 24$ sq. units

For any similar figures with corresponding sides of lengths L_1, L_2 respectively, and areas A_1 and A_2, then $\frac{A_1}{A_2} = \left(\frac{L_1}{L_2}\right)^2$.

The scale factor is $\left(\frac{L_1}{L_2}\right)$.

For volumes of similar solids, say V_1 and V_2, then $\frac{V_1}{V_2} = \left(\frac{L_1}{L_2}\right)^3$.

Combination of transformations

Two translations can be performed on a point (or a figure) by carrying out the operations separately or by combining the two translations into a single translation.

Example: A is the point (3, 1). Find the image after applying a translation of $\begin{pmatrix}-2\\5\end{pmatrix}$ followed by a further translation of $\begin{pmatrix}4\\-2\end{pmatrix}$.

In fig. (i), B is the image after first translation,
C is the image of B after second translation,
i.e. C is (5, 4).

In fig. (ii), we begin $\begin{pmatrix}-2\\5\end{pmatrix} + \begin{pmatrix}4\\-2\end{pmatrix} = \begin{pmatrix}-2 + 4\\5 + -2\end{pmatrix} = \begin{pmatrix}2\\3\end{pmatrix}$;

then applying translation of $\begin{pmatrix}2\\3\end{pmatrix}$ to A, we get D (5, 4).

Note C, D are the same position.

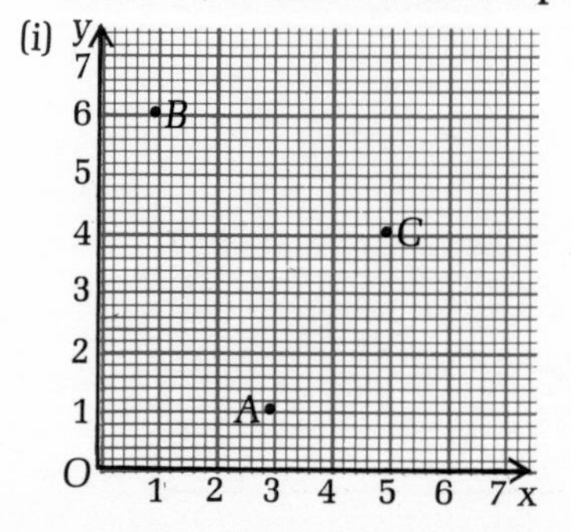

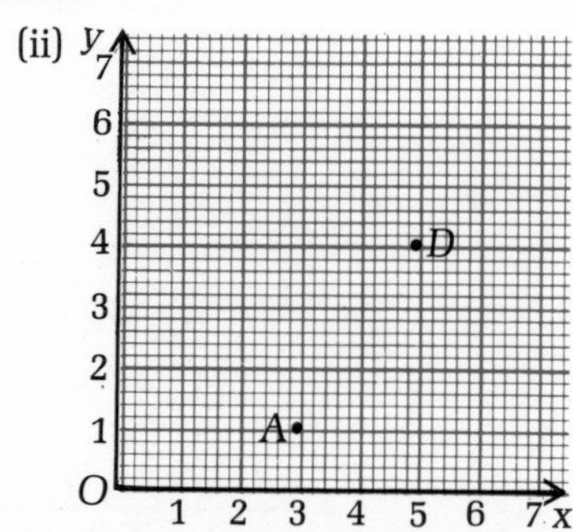

Combination of two reflections

We shall be concerned only with parallel mirror lines or mirror lines at 90° to each other.

Parallel mirror lines

In the diagram, the figure ⅂ is reflected in PQ and then in a second line of reflection RS. In the first reflection, the image is said to be *oppositely congruent* to the given figure. The second reflection produces a *directly congruent* image. The overall effect is to form a translation of the first reflection perpendicular to the mirror lines through twice the distance between the mirror lines.

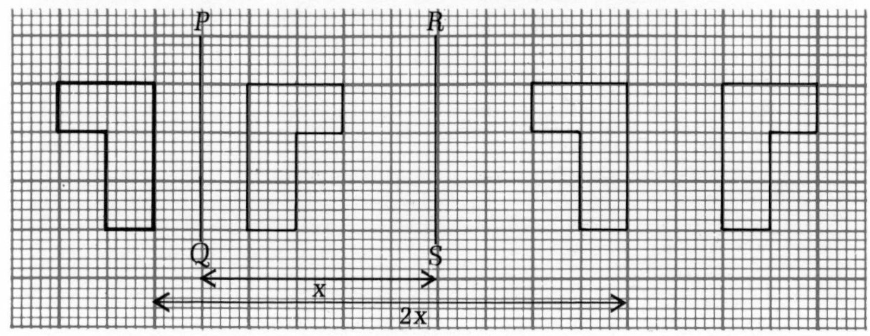

Mirror lines at 90° to each other

For reflections of an object in two perpendicular mirror lines of reflection, the effect is a rotation of −180° about the intersection of the mirror lines of reflection.

In the diagram, consider the axes to be the lines of reflection intersecting at O. Then B is the image of A for a reflection in the y-axis. C is a reflection of B in the x-axis. The effect is for C to be a rotation of A −180° about O.

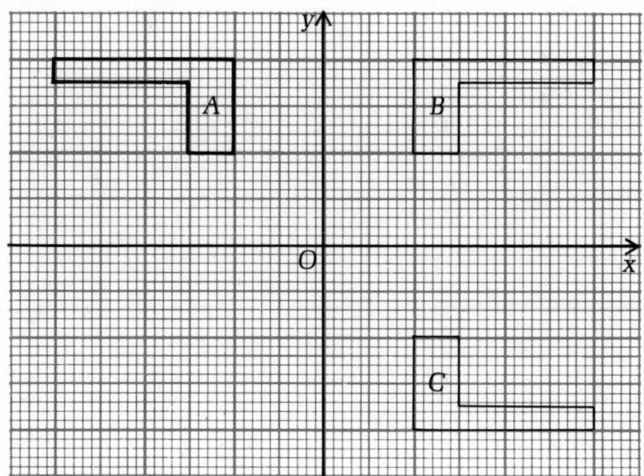

Two rotations

All rotations will be multiples of 90° and these were covered earlier. In effect, a rotation of an angle α followed by an angle β about some centre, if they are in the same sense, is a rotation of an angle $\alpha + \beta$ or $\alpha - \beta$ if they are in the opposite sense. Remember, both α and β will be some multiple of 90°.

In the diagram, the triangle *A* is rotated through 90° about the origin. Its image is *B*. The image of *B* is *C* after a rotation of 180° about *O*. The overall effect is a rotation of 270° about *O*, or 90° in the opposite sense.

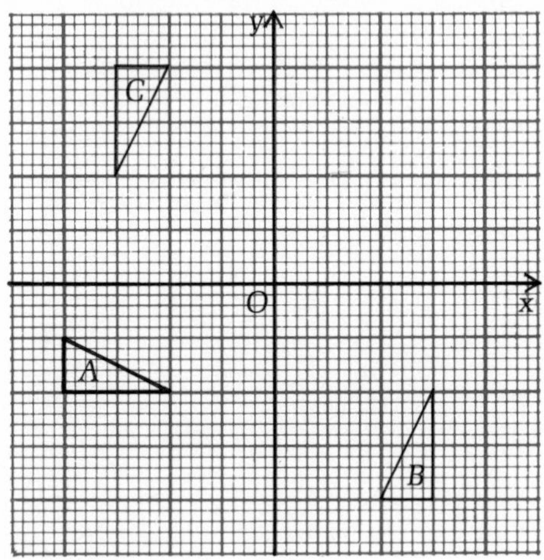

A rotation and a reflection

Again, rotations will be through some angle which is a multiple of 90°.

In the diagram, the figure *A* is rotated 180° about the origin *O* to get the image *B*. *B* is then reflected in line *MN* to get the image *C*. The effect is the same as translating *A* to *D* and reflecting the image in the *x*-axis. When such a translation is executed, it is called a glide.

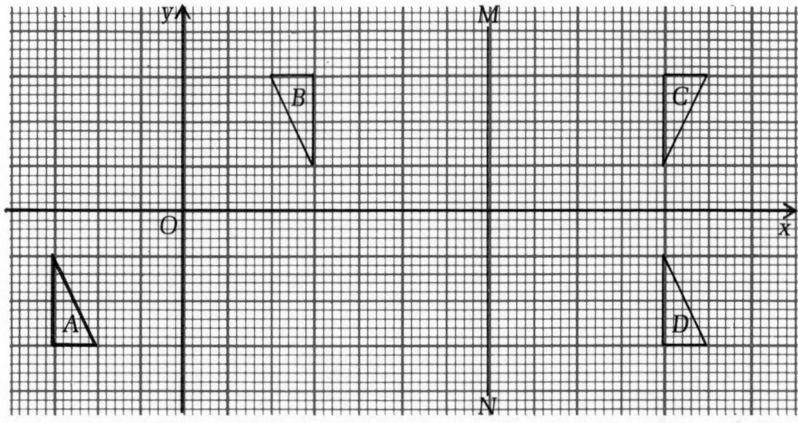

0 Matrices

Higher Level

Matrices, addition and subtraction of matrices, scalar multiplication, multiplication of two matrices, zero matrix, identity matrix, inverse of a 2×2 matrix.

Higher Level

Matrices

A matrix is an ordered array of numbers which acts as a store of information. For example, in one particular year the Premier Division of the Scottish Football League was as follows for the first five teams:

	Won	Drawn	Lost
Celtic	31	10	3
Hearts	23	16	5
Rangers	26	8	10
Aberdeen	21	17	6
Dundee United	16	15	13

Given the order of the teams in rows and the results in columns, we can represent the information as shown.

Each value of a matrix is called an *element*.

We use curved brackets to represent the matrix.

$$\begin{pmatrix} 31 & 10 & 3 \\ 23 & 16 & 5 \\ 26 & 8 & 10 \\ 21 & 17 & 6 \\ 16 & 15 & 13 \end{pmatrix}$$

A matrix is defined by its rows and columns, *i.e.* an $m \times n$ matrix has m rows and n columns. In the above example we have a 5×3 matrix. We always put the number of rows first, followed by the number of columns: this gives the *order* of the matrix.

In the examination, you might be presented with information to be put into matrix form.

Addition and subtraction of matrices

These operations are only possible if the matrices are of the same order.

Example 1: $\begin{pmatrix} 3 & 2 & 0 \\ 6 & 5 & 1 \end{pmatrix} + \begin{pmatrix} 2 & -6 & 7 \\ 1 & 1 & 2 \end{pmatrix} = \begin{pmatrix} 5 & -4 & 7 \\ 7 & 6 & 3 \end{pmatrix}$

Simply add the corresponding elements of each matrix.

Example 2: $\begin{pmatrix} 2 & 3 \\ 1 & -1 \end{pmatrix} - \begin{pmatrix} -3 & 4 \\ 0 & 1 \end{pmatrix} = \begin{pmatrix} 5 & -1 \\ 1 & -2 \end{pmatrix}$

In subtraction, subtract corresponding elements.

Subtraction of the following is impossible, since the orders of the two matrices are different:

$$\begin{pmatrix} 5 & 1 \\ 6 & 0 \end{pmatrix} - \begin{pmatrix} 3 & 1 & 6 \\ 1 & 1 & 1 \end{pmatrix}.$$

Scalar multiplication

A matrix multiplied by a single number means that each element is multiplied by that number.

Example: $5\begin{pmatrix} 3 & 6 \\ -1 & 0 \end{pmatrix} = \begin{pmatrix} 15 & 30 \\ -5 & 0 \end{pmatrix}$

Multiplication of two matrices

Two matrices are said to be comformable (or compatible) for multiplication if the number of columns of the first matrix is the same as the number of rows of the second matrix.

In general, it is better to write down the order of the matrices as follows:

If matrix A is of order $m_1 \times n_1$ and matrix B is of order $m_2 \times n_2$,

then $A \times B$

is $m_1 \times n_1$ $\quad m_2 \times n_2$

$n_1 = m_2$

Since $n_1 = m_2$, matrix multiplication is possible and product is of order $m_1 \times n_2$.

If $n_1 \neq m_2$, then product AB is impossible and matrices are said to be non-conformable for multiplication.

Example: If $A = \begin{pmatrix} 1 & 2 \\ 3 & 4 \end{pmatrix}$ and $B = \begin{pmatrix} 2 & 5 & 6 \\ 7 & 1 & 8 \end{pmatrix}$,

then $A \times B$

is 2×2 $\quad 2 \times 3$

conformable

order of AB is 2×3

NOTE: $B \times A$

2×3 $\quad 2 \times 2$

not conformable

$\therefore$ there is no matrix product BA.

In normal multiplication $AB = BA$, but matrix products are not commutative.

$$A \times B = \begin{pmatrix} 1 & 2 \\ 3 & 4 \end{pmatrix} \times \begin{pmatrix} 2 & 5 & 6 \\ 7 & 1 & 8 \end{pmatrix}$$

$$= \begin{pmatrix} 1 \times 2 + 2 \times 7 & 1 \times 5 + 2 \times 1 & 1 \times 6 + 2 \times 8 \\ 3 \times 2 + 4 \times 7 & 3 \times 5 + 4 \times 1 & 3 \times 6 + 4 \times 8 \end{pmatrix}$$

$$= \begin{pmatrix} 2 + 14 & 5 + 2 & 6 + 16 \\ 6 + 28 & 15 + 4 & 18 + 32 \end{pmatrix} = \begin{pmatrix} 16 & 7 & 22 \\ 34 & 19 & 50 \end{pmatrix}$$

In general: $\begin{pmatrix} a & b \\ c & d \end{pmatrix} \times \begin{pmatrix} p & q \\ r & s \end{pmatrix} = \begin{pmatrix} ap + br & aq + bs \\ cp + dr & cq + ds \end{pmatrix}$.

Zero matrix

This is a matrix with all its elements zero.

Example: $\begin{pmatrix} 0 & 0 \\ 0 & 0 \end{pmatrix}$

Identity matrix

An identity matrix is a square matrix (*i.e.* a matrix with the same number of rows as columns) with 1's on the leading diagonal going from the top left to the bottom right, all other elements being zeros.

Examples: $\begin{pmatrix} 1 & 0 \\ 0 & 1 \end{pmatrix}$ $\begin{pmatrix} 1 & 0 & 0 \\ 0 & 1 & 0 \\ 0 & 0 & 1 \end{pmatrix}$ $\begin{pmatrix} 1 & 0 & 0 & 0 \\ 0 & 1 & 0 & 0 \\ 0 & 0 & 1 & 0 \\ 0 & 0 & 0 & 1 \end{pmatrix}$

NOTE: $\begin{pmatrix} 1 & 0 \\ 0 & 1 \end{pmatrix} \times \begin{pmatrix} a \\ b \end{pmatrix} = \begin{pmatrix} 1 \times a + 0 \times b \\ 0 \times a + 1 \times b \end{pmatrix} = \begin{pmatrix} a \\ b \end{pmatrix}$

Multiplication by the Identity matrix (usually represented by *I*) on another matrix, say *A*, yields *A*, *i.e.* it is neutral for multiplication.

Inverse of a 2 × 2 matrix

If the product of two matrices, *A* and *B*, yields the Identity matrix *I*, *i.e.* $A \times B = I$ and $B \times A = I$, then *B* is called the inverse of *A* and is normally written as A^{-1}.

$$\therefore AA^{-1} = I$$

If $A = \begin{pmatrix} a & b \\ c & d \end{pmatrix}$, then $A^{-1} = \frac{1}{ad - bc} \begin{pmatrix} d & -b \\ -c & a \end{pmatrix}$

provided $ad - bc \neq 0$

If $ad - bc = 0$, the matrix $\begin{pmatrix} a & b \\ c & d \end{pmatrix}$ is said to be *singular.*

If $ad - bc \neq 0$, the matrix $\begin{pmatrix} a & b \\ c & d \end{pmatrix}$ is said to be *non-singular.* You will be concerned only with non-singular matrices.

The result $ad - bc$ is called the *determinant* of the matrix $\begin{pmatrix} a & b \\ c & d \end{pmatrix}$ and is written as $\begin{vmatrix} a & b \\ c & d \end{vmatrix}$.

A determinant is often written in shorthand form as Δ.

$\therefore$ If A^{-1} is the inverse of $A = \begin{pmatrix} a & b \\ c & d \end{pmatrix}$ to find A^{-1} then

$$A^{-1} = \frac{1}{\Delta} \begin{pmatrix} d & -b \\ -c & a \end{pmatrix} \text{ where } \Delta = \begin{vmatrix} a & b \\ c & d \end{vmatrix}$$

$\therefore$ the following stages would give the inverse A^{-1} of matrix $A = \begin{pmatrix} a & b \\ c & d \end{pmatrix}$.

Stage 1: Write the elements of A in determinant form $\begin{vmatrix} a & b \\ c & d \end{vmatrix}$

$\Delta = ad - bc$, i.e. product of elements on leading diagonal less product of elements on other diagonal.

Stage 2: Interchange elements on leading diagonal and change the signs of the elements on other diagonal.

Stage 3: $A^{-1} = \frac{1}{\Delta}\begin{pmatrix} d & -b \\ -c & a \end{pmatrix}$

Example: Find the inverse of $A = \begin{pmatrix} 5 & 2 \\ 3 & 4 \end{pmatrix}$.

$$\Delta = \begin{vmatrix} 5 & 2 \\ 3 & 4 \end{vmatrix} = 5 \times 4 - 2 \times 3 = 20 - 6 = 14$$

$$A^{-1} = \frac{1}{14}\begin{pmatrix} 4 & -2 \\ -3 & 5 \end{pmatrix}$$

$$\text{Check: } AA^{-1} = \begin{pmatrix} 5 & 2 \\ 3 & 4 \end{pmatrix} \times \frac{1}{14}\begin{pmatrix} 4 & -2 \\ -3 & 5 \end{pmatrix}$$

$$= \frac{1}{14}\begin{pmatrix} 5 & 2 \\ 3 & 4 \end{pmatrix}\begin{pmatrix} 4 & -2 \\ -3 & 5 \end{pmatrix}$$

$$= \frac{1}{14}\begin{pmatrix} 20 - 6 & -10 + 10 \\ 12 - 12 & -6 + 20 \end{pmatrix}$$

$$= \frac{1}{14}\begin{pmatrix} 14 & 0 \\ 0 & 14 \end{pmatrix} \qquad \left[\text{or } \frac{1}{14} \times 14 \begin{pmatrix} 1 & 0 \\ 0 & 1 \end{pmatrix}\right]$$

$$= \begin{pmatrix} 1 & 0 \\ 0 & 1 \end{pmatrix}$$

$$\text{or } A^{-1}A = \frac{1}{14}\begin{pmatrix} 4 & -2 \\ -3 & 5 \end{pmatrix}\begin{pmatrix} 5 & 3 \\ 3 & 4 \end{pmatrix}$$

$$= \frac{1}{14}\begin{pmatrix} 20 - 6 & 8 - 8 \\ -15 + 15 & -6 + 20 \end{pmatrix}$$

$$= \frac{1}{14}\begin{pmatrix} 14 & 0 \\ 0 & 14 \end{pmatrix}$$

$$= \begin{pmatrix} 1 & 0 \\ 0 & 1 \end{pmatrix}$$

Vectors

Intermediate Level
Vectors, position vector, addition of vectors, multiplication by a scalar, modulus, equal vectors.

Intermediate Level

Vectors

A vector is a quantity which has both magnitude and direction. For instance, an aeroplane flying at 800 km/h on a bearing of 160° from Manchester can be represented by a vector.

A translation, mentioned in module 19, can be defined in terms of a column vector.

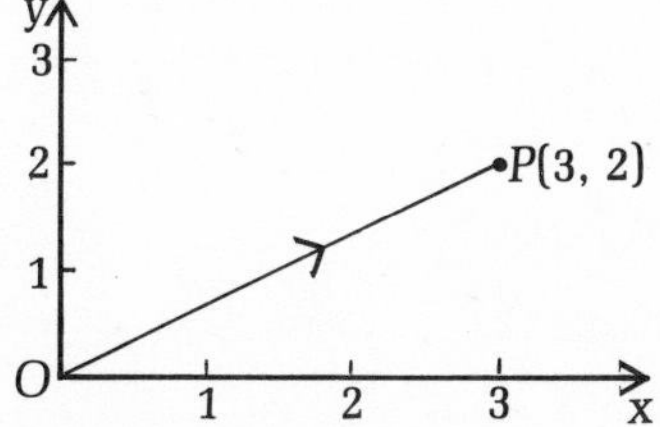

If *OP* is a vector, then the length of *OP* represents the magnitude and the arrow represents the direction. The vector *OP* is usually written as $\overrightarrow{OP}$ or **OP** or $\underset{\sim}{OP}$. Sometimes this vector is written as **p** in bold print or $\underset{\sim}{p}$.

The shift (translation) from *O* to *P* is written as $\begin{pmatrix}3\\2\end{pmatrix}$ in order to distinguish it from the point *P* (3, 2). The column vector $\begin{pmatrix}3\\2\end{pmatrix}$ means a shift of 3 units to the right followed by a shift of 2 units upwards.

Example 1: What translation takes (3, 5) to (7, 2)?

Let the shift vector be $\begin{pmatrix}x\\y\end{pmatrix}$.

$\begin{pmatrix}3\\5\end{pmatrix} + \begin{pmatrix}x\\y\end{pmatrix} = \begin{pmatrix}7\\2\end{pmatrix}$ i.e. $\begin{pmatrix}x\\y\end{pmatrix} = \begin{pmatrix}4\\-3\end{pmatrix}$

Example 2: To what point does the translation with shift vector $\begin{pmatrix}2\\3\end{pmatrix}$ take the point (4, 2)?

$\begin{pmatrix}4\\2\end{pmatrix} + \begin{pmatrix}2\\3\end{pmatrix} = \begin{pmatrix}6\\5\end{pmatrix}$

The new point is (6, 5).

Position vector

The position vector of a point is the shift vector of the translation which takes the origin to that point. Hence the position vector of any point (x, y) is $\begin{pmatrix}x\\y\end{pmatrix}$.

Magnitude of a vector

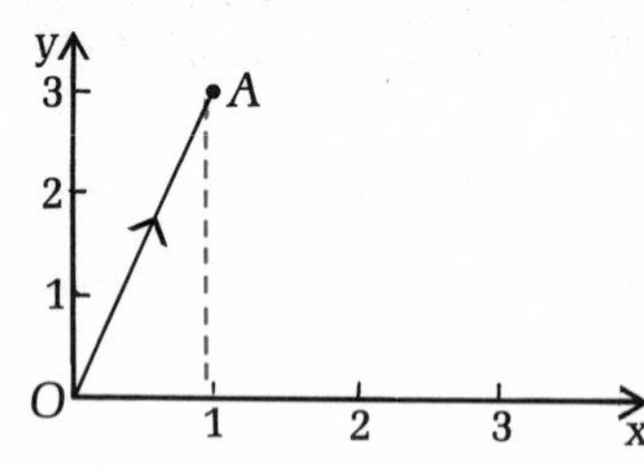

The magnitude of $\overrightarrow{OA}$ is the length of OA and is found by using Pythagoras's Theorem.

$$OA^2 = 1^2 + 3^2 = 1 + 9 = 10$$

$$\therefore \text{ magnitude of } \overrightarrow{OA} = \sqrt{10}$$

In general:

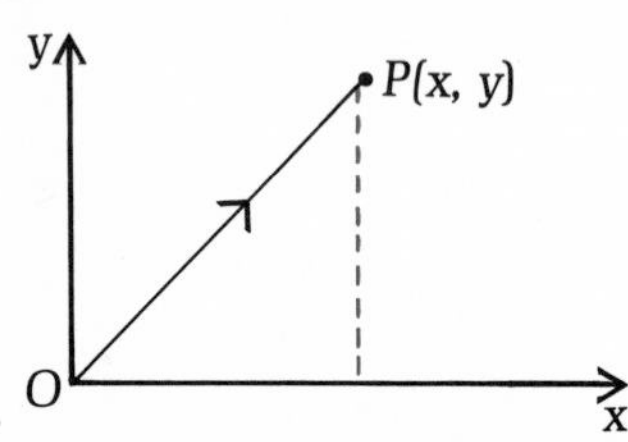

$$\text{magnitude of } \overrightarrow{OP} = \sqrt{x^2 + y^2}.$$

Addition of vectors

Suppose we have two points, A (1, 3) and B (4, 1), plotted on the graph shown below.

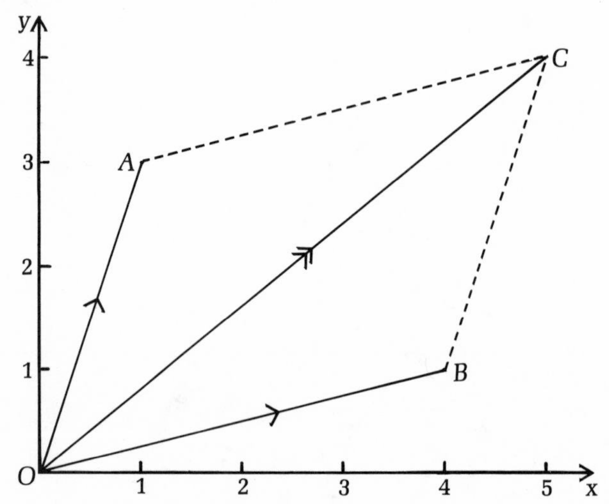

Then the position vector of A is $\begin{pmatrix}1\\3\end{pmatrix}$ and the position vector of B is $\begin{pmatrix}4\\1\end{pmatrix}$.

Then $\overrightarrow{OA} + \overrightarrow{OB} = \begin{pmatrix}1\\3\end{pmatrix} + \begin{pmatrix}4\\1\end{pmatrix} = \begin{pmatrix}5\\4\end{pmatrix}$,

i.e. the position vector of another point (5, 4) which we will call C. Note that C is fixed by completing the parallelogram $OACB$.

We write $\overrightarrow{OA} + \overrightarrow{OB} = \overrightarrow{OC}$

or if **a** is the position vector of a point A
and **b** is the position vector of a point B
and **c** is the position vector of a point C
then $\mathbf{a} + \mathbf{b} = \mathbf{c}$.

Note also that in a parallelogram, since opposite sides are parallel (same direction) and equal in length,

then $\quad \overrightarrow{OA} = \overrightarrow{BC}$

$\therefore \overrightarrow{OB} + \overrightarrow{OA} = \overrightarrow{OB} + \overrightarrow{BC} = \overrightarrow{OC}$

or $\quad \overrightarrow{OB} = \overrightarrow{AC}$

$\therefore \overrightarrow{OA} + \overrightarrow{OB} = \overrightarrow{OA} + \overrightarrow{AC} = \overrightarrow{OC}$.

In general: $\overrightarrow{AB} + \overrightarrow{BC} + \overrightarrow{CD} + \overrightarrow{DE} = \overrightarrow{AE}$.

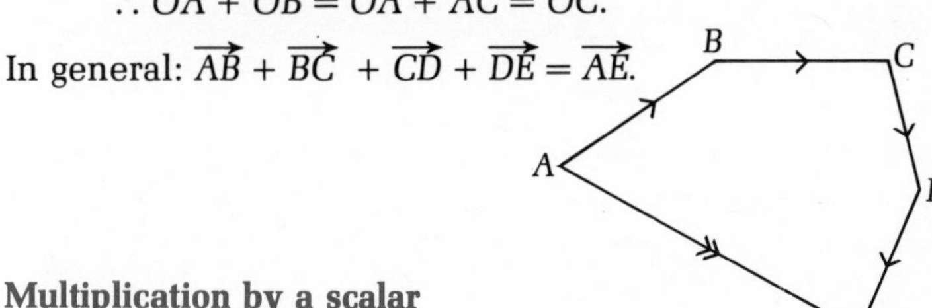

Multiplication by a scalar

If $\overrightarrow{OP} = \begin{pmatrix} 2 \\ 3 \end{pmatrix}$,

then $2\,\overrightarrow{OP} = \begin{pmatrix} 4 \\ 6 \end{pmatrix}$

$= \overrightarrow{OQ}$.

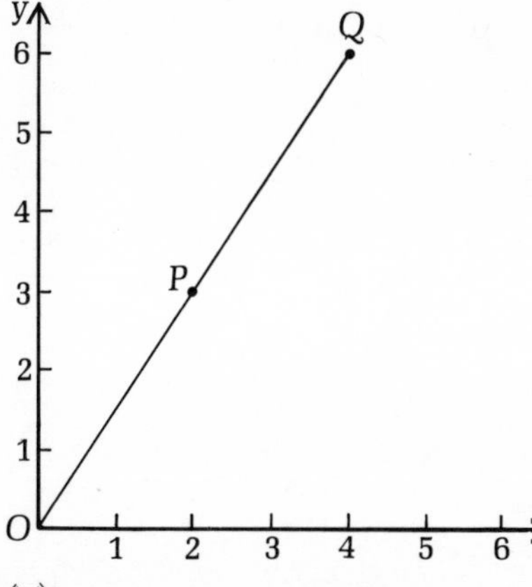

OQ is twice as long as OP.

OPQ is a straight line.

In general: if $\overrightarrow{OP} = \begin{pmatrix} x \\ y \end{pmatrix}$ and n is any number,

$$n\,\overrightarrow{OP} = \begin{pmatrix} nx \\ ny \end{pmatrix}.$$

Modulus

The magnitude of a vector $\begin{pmatrix} x \\ y \end{pmatrix}$ is $\sqrt{x^2 + y^2}$

The magnitude of any vector **a** is known as its modulus and is denoted by

$$|\mathbf{a}| \quad \text{or} \quad a.$$

The modulus is a scalar quantity and is always positive, except in the obvious case of the zero vector.

Equal vectors

Vectors are said to be equal if they have the same magnitude and the same direction:

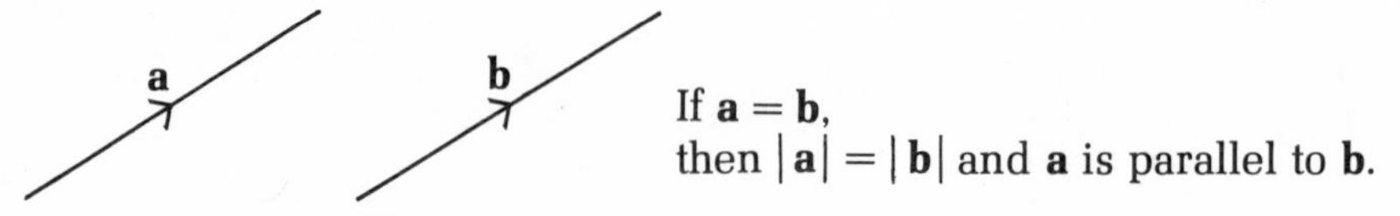

If $\mathbf{a} = \mathbf{b}$,
then $|\mathbf{a}| = |\mathbf{b}|$ and **a** is parallel to **b**.

If the two vectors have the same magnitude but are in opposite directions

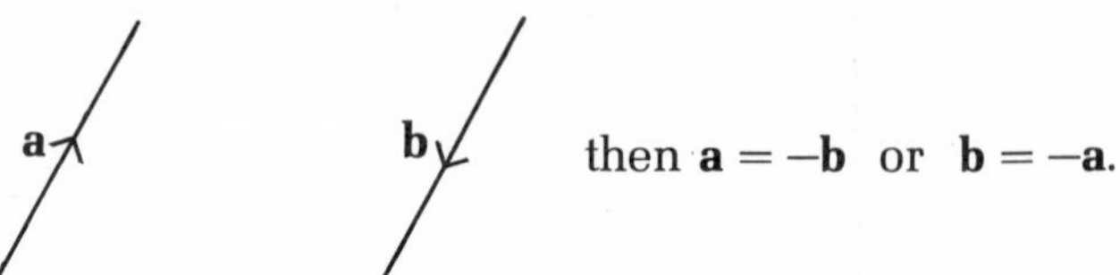

then $\mathbf{a} = -\mathbf{b}$ or $\mathbf{b} = -\mathbf{a}$.

It follows that if we have two vectors, say **a** and 3**a**, they are in the same direction but one is represented by a length three times the other:

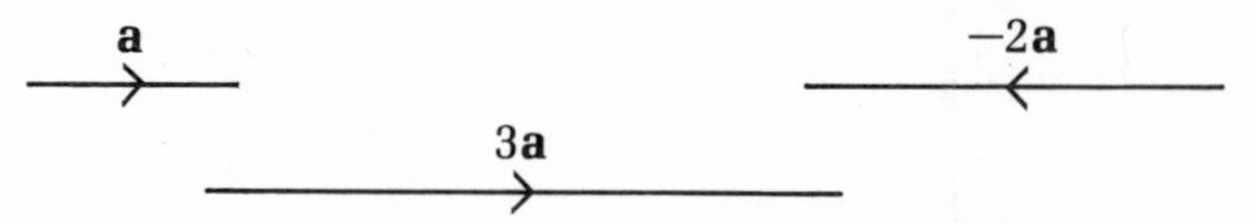

The vectors are represented by directed line segments. Vectors can then be used to establish simple geometrical results.

Example: Show that the line joining the midpoints of two sides of a triangle is parallel and equal to half the third side of the triangle.

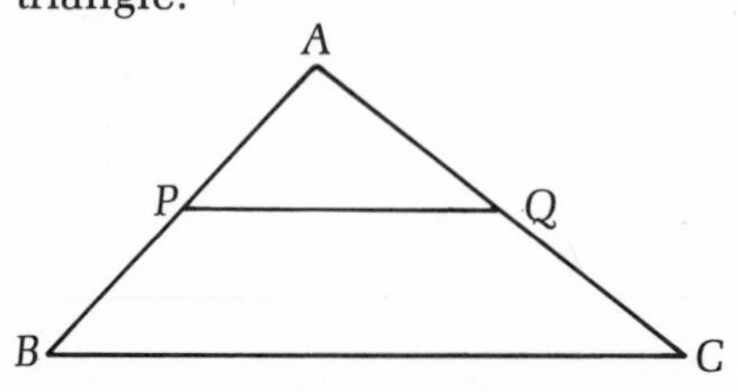

Let P and Q be midpoints of AB and AC in a triangle ABC.

Let $\overrightarrow{AB}$ and $\overrightarrow{AC}$ be **a** and **b**

then $\overrightarrow{BC} = \overrightarrow{BA} + \overrightarrow{AC} = -\overrightarrow{AB} + \overrightarrow{AC} = -\mathbf{a} + \mathbf{b}$

$\overrightarrow{AP} = \frac{1}{2}\mathbf{a}$ $\qquad$ $\overrightarrow{AQ} = \frac{1}{2}\mathbf{b}$

then $\overrightarrow{PQ} = \overrightarrow{PA} + \overrightarrow{AQ} = -\overrightarrow{AP} + \overrightarrow{AQ} = -\frac{1}{2}\mathbf{a} + \frac{1}{2}\mathbf{b} = \frac{1}{2}(-\mathbf{a} + \mathbf{b})$

hence length of PQ is $\frac{1}{2}$ of BC and PQ is parallel to BC since $\overrightarrow{PQ} = \frac{1}{2}\overrightarrow{BC}$.